PEDIATRIC EMERGENCY MEDICINE BOARD REVIEW

THIRD EDITION

PEDIATRIC EMERGENCY MEDICINE BOARD REVIEW

THIRD EDITION

William Gossman, MD, FAAEM
Jason Langenfeld, MD, FACEP
Scott Plantz, MD, FAEEM

PEDIATRIC EMERGENCY MEDICINE BOARD REVIEW

Pediatric Emergency Medicine Board Review, Third Edition

Copyright © 2017 by Wenckebach Publishing, Chicago, IL. All rights reserved, including the right of reproduction, in whole or in part, in any form. Printed in the United States of America. Except as permitted under the United States Copyright Act of 1976, no part of this publication may be reproduced or distributed in any form or by any means, or stored in a data base or retrieval system, without the prior written permission of the publisher.

ISBN 978-1-365-93431-5

Notice
Medicine is an ever-changing science. As new research and clinical experience broaden our knowledge, changes in treatment and drug therapy are required. The authors and the publisher of this work have checked with sources believed to be reliable in their efforts to provide information that is complete and generally in accord with the standards accepted at the time of publication. However, in view of the possibility of human error or changes in medical sciences, neither the authors nor the publisher nor any other party who has been involved in the preparation or publication of this work warrants that the information contained herein is in every respect accurate or complete, and they disclaim all responsibility for any errors or omissions or for the results obtained from use of the information contained in this work. Readers are encouraged to confirm the information contained herein with other sources.

Layout and Cover Design - Sherry Gossman, RN, BSN

DEDICATION

To my wife Sherry - William Gossman

To Angela, for your love, support, and tolerance - Jason Langenfeld

To my wife Cynna - Scott Plantz

AUTHORS AND EDITORS

EDITORS:

William Gossman M.D. FAAEM
Chairman of Emergency Medicine
Creighton University Medical School
Omaha, NE

Jason Langenfeld, M.D. FACEP
Assistant Professor
Department of Emergency Medicine
University of Nebraska Medical Center
Omaha, NE

Scott H. Plantz, M.D. FAAEM
Professor of Emergency Medicine
University of Louisville
Louisville, KY

CONTRIBUTING AUTHORS:

Amy Archer, MD, FACEP
Attending Physician
Advocate Lutheran General Hospital
Department of Emergency Medicine

Tracy Boykin M.D. MBA, MPH, FAAEM
Assistant Professor
Department of Emergency Medicine
Loyola University, Stritch School of Medicine
Maywood, IL

William U. Britton, M.D.
Department of Emergency Medicine
University of Nebraska Medical Center
Omaha, NE

June A. Burn, M.D.
Department of Emergency Medicine
University of Nebraska Medical Center
Omaha, NE

George Chiampas, DO, FACEP
Clinical Assistant Professor
Northwestern University Feinberg School of Medicine,
Department of Emergency Medicine

Steve C. Christos, D.O. FACEP, FAAEM
Clinical Assistant Professor
Presence Resurrection Medical Center
Department of Emergency Medicine

Jeffrey Cooper, M.D.
Associate Professor
Department of Emergency Medicine
University of Nebraska Medical Center
Omaha, NE

Nicole Colucci, DO, FACEP, FAAP
Clinical Assistant Professor
Presence Resurrection Medical Center
Department of Emergency Medicine
Loyola University Medical Center
Department of Surgery, Section of Emergency Medicine

Brian Donahue, MD, FACEP
Clinical Assistant Professor
Presence Resurrection Medical Center
Department of Emergency Medicine

Eric V. Ernest, M.D. EMT-P, FACEP, FAEMS
Assistant Professor
Department of Emergency Medicine
University of Nebraska Medical Center
Omaha, NE

Mitchell J. Goldman, D.O. FAAEM, FAAP
Director, Pediatric Emergency Medicine
St. Vincent Emergency Physician, Inc.
Indianapolis, IN

Charles Justin Hall D.O.
Department of Emergency Medicine
University of Nebraska Medical Center
Omaha, NE

Cynthia Hernandez, M.D.
Assistant Professor
Department of Emergency Medicine
University of Nebraska Medical Center
Omaha, NE

Saleem Jafilan, M.D.
Department of Emergency Medicine
University of Nebraska Medical Center
Omaha, NE

Matthew Jordan, MD, FACEP
Clinical Assistant Professor
Presence Resurrection Medical Center
Department of Emergency Medicine

PEDIATRIC EMERGENCY MEDICINE BOARD REVIEW

Stephen W. Leslie M.D. FACS
Associate Professor of Surgery
Division of Urology
Creighton University School of Medicine
Omaha, NE

Ross Mathiasen, M.D.
Assistant Professor
Department of Emergency Medicine
Department of Orthopaedic Surgery
University of Nebraska Medical Center
Omaha, NE

William Parente, MD
Emergency Medicine Resident
Presence Resurrection Medical Center
Department of Emergency Medicine

Mark Postel, DO
Emergency Medicine Resident
Presence Resurrection Medical Center
Department of Emergency Medicine

Thomas J. Rankin, M.D. FACEP
Staff Physician
Methodist Health System
Omaha, NE

Nichole M. Schafer, MD
Department of Emergency Medicine
University of Nebraska Medical Center
Omaha, NE

James L Shoemaker Jr., MD, FACEP
Assistant Professor Department of Emergency Medicine
Elkhart Emergency Physicians Inc.
EEPI Director of Quality
Immediate Past President, IN Chapter ACEP

CONTRIBUTORS TO PREVIOUS EDITIONS

Bobby Abrams, M.D.
Jonathan Adler, M.D.
Raymond C. Baker, M.D.
Kristen Bechtel, M.D., MCP
Michelle L. Bez, D.O.
Frank M. Biro, M.D.
David F.M. Brown, M.D.
Sharona D. Bryant, M.D.
Philadelphia, PA
Eduardo Castro, M.D.
Deandra Clark, M.D.
David Cone, M.D.
Cheryl A. Coyle, RN, BSN, CCRN

Carl W. Decker, M.D.
Phillip G. Fairweather, M.D.
Craig Feied, M.D.
David Neal Franz, M.D.
Lynn Garfunkel, M.D.
Marianne Gausche-Hill, MD, FACEP,
Jay Gold, M.D.
Javier A. Gonzalez del Rey, M.D.
John Graneto, D.O.
Michael I. Greenberg, M.D., MPH
Guy H. Haskell, Ph.D., NREMT-P
Ned Hayes, M.D.
Robert G. Hendrickson, M.D.
Melanie S. Heniff, MD, FAAEM, FAAP
Rita Herrington-Mikula, BSN, RN, EMT-P
James F. Holmes, M.D.
Eddie Hooker, M.D.
Ira Horowitz, M.D.
Matt Kopp, M.D.
Lieutenant Robert C. Krause, EMT-P
Lance W. Kreplick, M.D.
Deborah A. Lee, M.D., Ph.D.
Robert M. Levin, M.D.
Gillian Lewke Emblad, PA, CMA
Bernard Lopez, M.D.
Mary Nan S. Mallory, M.D.
Mariann M. Manno, M.D., FAAP
Bridget A. Martell, M.D.
David Morgan, M.D.
Anthony Morocco, M.D.
James J. Nordlund, M.D.
Peter Noronha, M.D.
cott E. Olitsky, M.D.
Robert L. Palmer, PA-C, REMT-P
Geraldo Reyes, M.D.
Luis R. Rodriquez, M.D., FAAP
Carlo Rosen, M.D.
Bruce K. Rubin, M.D.
Howard M. Saal, M.D.
Rejesh Shenoy, M.D.
Mahesh Shrestha, M.D.
Clifford S. Spanierman, M.D., FAAP
Dana Stearns, M.D.
Jack Stump, M.D.
Loice Swischer, M.D.
Nicholas Tapas, M.D.
Hector Trujillo, M.D.
William T. Zempsky, M.D.
Michael Zevitz, M.D.
Huiquan Zhao, M.D., Ph.D.
Stan Zuba M.D.

INTRODUCTION

Congratulations!

Pediatric Emergency Medicine Board Review, will help you learn some medicine. Originally designed as a study aid to improve performance on the Emergency Pediatric Written Board exams or Emergency Pediatric In-service exam, this book is full of useful information and a few words are appropriate when discussing intent, format, limitations, and use.

Since Pediatric Emergency Medicine Board Review is primarily intended as a study aid, the text is written in rapid-fire question/answer format. This way, readers receive immediate gratification without the misleading or confusing "foils". This eliminates the risk of erroneously assimilating an incorrect piece of information that makes a big impression. Questions themselves often contain a "pearl" intended to reinforce the answer. Additional "hooks" may be attached to the answer in various forms, including mnemonics, visual imagery, repetition, and humor. Additional information not requested in the question may be included in the answer. Emphasis has been placed on distilling trivia and key facts that are easily overlooked, that are quickly forgotten, and that somehow seem to be needed on board examinations.

Many questions have answers without explanations. This enhances ease of reading and rate of learning. Explanations often occur in a later question/answer. Upon reading an answer, the reader may think, "Hm, why is that?" or, "Are you sure?" If this happens to you, go check! Truly assimilating these disparate facts into a framework of knowledge absolutely requires further reading of the surrounding concepts. Information learned in response to seeking an answer to a particular question is retained much better than information that is passively observed. Take advantage of this! Use this book with your preferred source texts handy and open.

Pediatric Emergency Medicine Board Review has limitations. We have found many conflicts between sources of information. We have tried to verify in several references the most accurate information. Some texts have internal discrepancies further confounding clarification.

Pediatric Emergency Medicine Board Review risks accuracy by aggressively pruning complex concepts down to the simplest kernel—the dynamic knowledge base and clinical practice of medicine is not like that! Furthermore, new research and practice occasionally deviates from that which likely represents the right answer for test purposes. This text is designed to maximize your score on a test. Refer to your most current sources of information and mentors for direction.

We welcome your comments, suggestions and criticism. Great effort has been made to verify these questions and answers. Some answers may not be the answer you would prefer. Most often this is attributable to variance between original sources. Please make us aware of any errors you find. We hope to make continuous improvements and would greatly appreciate any input with regard to format, organization, content, presentation, or about specific questions. We also are interested in recruiting new contributing authors and publishing new textbooks. We look forward to hearing from you!

Study hard and good luck!

W.G., J.L., & S.P.

TABLE OF CONTENTS

EMERGENCY MEDICAL SERVICES 03
PEDIATRIC LIFE SUPPORT 07
INFECTIOUS DISEASES 55
CARDIOVASCULAR 87
SEDATION AND ANALGESIA 117
PULMONARY 123
GASTROINTESTINAL 139
TOXICOLOGY AND ENVIRONMENTAL 157
METABOLIC, ENDOCRINE AND NUTRITION 235
SUDDEN INFANT DEATH SYNDROME AND APPARENT LIFE-THREATENING
EVENT 247
NEUROLOGY 249
RHEUMATOLOGY, IMMUNOLOGY AND ALLERGY 263
GENITOURINARY AND RENAL 271
HEMATOLOGY AND ONCOLOGY 281
DERMATOLOGY 289
SURGICAL EMERGENCIES 295
TRAUMA 301
ORTHOPEDIC EMERGENCIES 317
EYES, EARS, NOSE, THROAT & NECK 327
PSYCHOSOCIAL AND CHILD ABUSE 335
OBSTETRICS AND GYNECOLOGY 343
LEGAL 355
BIBLIOGRAPHY 363

EMERGENCY MEDICAL SERVICES

☐ **At which three levels of government should EMS-C be integrated into EMS?**

It must be integrated at the state, regional, and local levels.

☐ **Which two agencies provide funding for education, training and research needed by EMS systems to treat pediatric patients?**

1. U.S. Dept. of Health and Human Services, Health Services and Resources Administration, Maternal and Child Health Branch.

2. The National Highway Traffic Safety Administration.

☐ **Identify the four links in pediatric Chain of Survival.**

Injury prevention, early CPR, early access to EMS, early advanced life support, and integrated Post-cardiac arrest care.

☐ **List the two components of the first link (injury prevention) in the Chain of Survival for pediatric emergencies.**

Identification of problems and early intervention.

☐ **When does entry into the EMS system occur?**

Entry occurs when the rescuer or bystander calls local EMS and speaks with a dispatcher.

☐ **EMS dispatchers often provide a caller with pre-arrival directions. Who dictates what information is given to the caller?**

The information given callers varies widely and is controlled by local jurisdiction and medical oversight.

☐ **Explain the advantages of using established dispatcher protocols when providing a caller with pre-arrival instructions.**

Use of dispatcher protocols will reduce the variability of information provided by dispatchers and ensure that succinct, accurate emergency information is provided to every caller in need of emergency assistance.

☐ **In the broad scope, "who" is actually the first link in the Chain of Survival?**

The first link in the Chain of Survival is the general public.

PEDIATRIC EMERGENCY MEDICINE BOARD REVIEW

☐ **Who is included in this "first link" as it applies to children?**

Parents, caretakers, older siblings, and school personnel, who must be educated about the proper use of the EMS system.

☐ **Approximately what percentage of EMS contacts in both rural and urban areas are children under 14 years of age?**

10%.

☐ **For children younger than five years of age, is the EMS system most frequently activated for trauma or medical illness?**

Medical illness.

☐ **Serious illness, including cardiopulmonary arrest, is most common in children of what age?**

2 years of age.

☐ **How can a child acquire status of an "emancipated minor"?**

Marriage, judicial decree, military service, parental consent, failure of parents to meet legal responsibilities, living apart from and being financially independent of parents, motherhood.

☐ **An EMT responds to a call where there is a question as to the authenticity of documents stating the victim's intent to refuse CPR. The victim is in cardiac arrest. What should the EMT do?**

The practice of initiating resuscitative efforts is generally accepted. CPR may be discontinued with further authentication of the patient's wishes or with a MD order.

☐ **T/F: The Good Samaritan law protects the EMT from acts of negligence.**

False. The traditional legal concept holds the EMT to a "reasonable care" standard. The circumstances surrounding the incident are evaluated and the skill level of the rescuer is taken into consideration.

☐ **Should the first responder attempt to assess resuscitatability of the cardiac arrest patient and then determine whether or not to initiate CPR?**

No. First responders are encouraged to initiate CPR in the cardiac arrest victim. Currently there are no accepted criteria for immediate determination of death by first responders. There are few exceptions to this rule.

☐ **When does the legal duty to meet a "reasonable standard" of care commence for the healthcare provider?**

Once he or she performs an act that may be construed as rendering care.

EMERGENCY MEDICAL SERVICES

☐ **What are "mature minor" rules?**

Statutory rules that uphold the validity of consent given by minors if the treatment is appropriate and the minor is considered capable of comprehending the clinical circumstances and therapeutic options.

☐ **T/F: Medical malpractice cases concerning CPR have been limited to in-hospital cases in which the risk of cardiac arrest could be anticipated.**

True. Successful action against laypersons that attempt out-of-hospital CPR seems unlikely.

☐ **On what does the paramedic's scope of practice rely?**

Standing order protocols and offline medical direction (indirect medical control).

☐ **Once you have initiated CPR, do you incur a legal liability to continue?**

Once steps are initiated and termination would place the victim in a worse position or compromise the likelihood of assistance from others, the potential rescuer incurs a legal duty to perform appropriately.

☐ **Do statutes cover events that take place during transport to a medical facility?**

Rescue settings vary by statute. In some states only actions at the site of the emergency are covered; other states include events during transport to a medical facility.

☐ **Why should first responders not fear litigation from making reasonable attempts at CPR?**

Because CPR attempts in emergencies are likely to be viewed as reasonable, even when provided by an inadequately trained rescuer, as long as there is a good-faith belief that the possible benefits of the attempt outweigh the risk from the rescuer's incompetence.

☐ **Must pre-hospital personnel honor hospital DNR orders?**

Many state laws still make it illegal for pre-hospital personnel to honor a hospital DNR order.

☐ **May an EMT disregard or countermand an order given by a physician?**

Not as a rule. However, most EMS systems have specific policies dealing with this issue. The procedure used varies widely from system to system.

☐ **At what age can an automated external defibrillator (AED) be used in children?**

Eight or older.

☐ **Why is hypothermia a particular concern in the field treatment of small children?**

Children have poorly developed heat regulation mechanisms (e.g. poor shivering), and a relatively larger surface area to volume ratio than adults. Accordingly, infants and small children are more likely to become hypothermic if exposed to the cold for prolonged periods, as may occur during extrication from a motor vehicle crash.

PEDIATRIC EMERGENCY MEDICINE BOARD REVIEW

- **Medics are called to a group home where a 15 y o male began smashing furniture. The medics arrive before law enforcement and the scene is not secure. What should the medics do?**

    EMS personal must protect themselves from harm, so they should wait for law enforcement to arrive and secure the scene.

- **A pediatrician has a child in the office that is having an asthma attack. The child does not improve after nebulizer treatment, what should the pediatrician do?**

    The pediatrician should call 911. It would be unsafe to transport the patient by private auto.

- **What is the only federal program whose purpose was to improve the quality of pediatric emergency care.**

    The Emergency Medicine Services for Children (EMSC) program.

- **Paramedics are called to the home of a 13 year old aggressive bipolar patient. The patient is smashing furniture and the police have yet to arrive. What should they do?**

    EMS personnel must protect themselves from harm so transport will occur once the police arrive.

PEDIATRIC LIFE SUPPORT

☐ **T/F: Primary cardiac arrest is the most common cause of death in young children.**

False. It is uncommon in children. Respiratory arrest usually precedes cardiac arrest.

☐ **What percent of children present with ventricular fibrillation as a cause of pulseless arrest outside of the hospital?**

10-15%. Most are in asystole.

☐ **Ventricular fibrillation is more likely in which patients?**

Children > 10, submersion victims, those with congenital heart disease and in-hospital arrest.

☐ **What is the common primary cause of pulseless cardiac arrest?**

Injury or disease leading to respiratory or circulatory failure.

☐ **Cardiopulmonary arrest occurs predominately in which two pediatric age groups?**

Infants < 1year, and adolescents.

☐ **What is the most common cause of death in children >1 year of age?**

Unintentional injury.

☐ **Motor vehicle trauma accounts for what percent of all pediatric injuries and deaths?**

Almost 50%.

☐ **What is the most important device to prevent pediatric motor vehicle injuries and deaths?**

A properly installed child restraint device.

☐ **What percent of head injuries could bike helmets eliminate?**

85% of head injuries and 88% of brain injuries.

☐ **What percent of fire-related deaths result from house fires?**

80%.

PEDIATRIC EMERGENCY MEDICINE BOARD REVIEW

☐ **T/F: Firearms are a leading cause of accidental death and injury in children.**

True.

☐ **T/F: The presence of a gun in the home is linked to adolescent suicide.**

True.

☐ **T/F: An awake child with respiratory distress should be laid on the ground or on a stretcher.**

False. Children with partial airway obstruction will find the most comfortable position on their own.

☐ **If the child is conscious, when should EMS be activated?**

Immediately.

☐ **If the child is unconscious, how long should the rescuer provide BLS before activating EMS?**

2 minutes. Most pediatric arrests are respiratory in origin, so two minutes may be enough to restore ventilation and oxygenation or prevent respiratory arrest from progressing to cardiac arrest.

☐ **In an unconscious child, when should EMS be immediately activated?**

When two rescuers are present.

☐ **What should be done if trauma is suspected?**

Complete immobilization of the spine.

☐ **How can the airway be opened in an unconscious, apneic victim?**

Head tilt chin-lift, and/or jaw thrust.

☐ **How can the airway be opened in an unconscious, apneic victim if neck trauma is suspected?**

Jaw thrust with neck immobilized.

☐ **How is the head tilt-chin lift performed?**

One hand is used to tilt head and gently extend the neck. The index finger of the opposite hand lifts the chin outward.

☐ **After the airway is established, how does the rescuer assess breathing?**

Looks for rise and fall of the chest, listens for exhaled air, feels for exhaled airflow at the mouth.

PEDIATRIC LIFE SUPPORT

☐ **What is the recovery position used for?**

When the victim is unconscious but has no evidence of trauma and is breathing effectively.

☐ **How is the victim moved into the recovery position?**

The head, torso, and shoulders are moved simultaneously so that the victim is on his or her side.

☐ **T/F: Rescue breathing is provided for all victims.**

False. Only those victims who have no spontaneous breathing.

☐ **In infants, how should rescue breaths be provided?**

By placing your mouth over the victim's nose and mouth.

☐ **How many breaths are supplied in rescue breathing?**

Two slow breaths, 1-1.5 seconds each separated by a pause to take a breath.

☐ **What is the purpose of the pause between rescue breaths?**

To maximize oxygen content.

☐ **How does the rescuer know if ventilation is not effective?**

The child's chest does not rise.

☐ **Why should breaths be delivered slowly?**

To minimize the pressure required for ventilation, and prevent gastric distension.

☐ **What is the most common cause of an airway obstruction?**

Laxity in the tongue causing it to fall back on the pharynx.

☐ **What should the rescuer do if ventilation is unsuccessful?**

Reposition the head in the midline, place a towel under the shoulders and consider progressively extending the neck (provided trauma is not suspected), make a better seal and try again.

☐ **What happens when cricoid pressure is administered?**

The trachea is displaced posteriorly compressing the esophagus against the vertebral column.

PEDIATRIC EMERGENCY MEDICINE BOARD REVIEW

☐ **In a patient with a tracheostomy, how is air leakage prevented during rescue breathing?**

The victim's mouth and nose are sealed by the rescuer's hand or a tight fitting mask.

☐ **When should the rescuer determine the need for chest compressions?**

If the child is unresponsive and not breathing, CPR should be started immediately.

☐ **What complications are common in children who receive CPR?**

None. Rib fractures in children are usually attributable to preceding trauma.

☐ **How long should the layperson spend looking for a pulse in a non-breathing child?**

No time. Laypersons often have difficulty recognizing pulses and should begin chest compressions immediately if the patient is not breathing or moving.

☐ **Where should the pulse be checked by health care professionals in infants?**

Brachial artery.

☐ **Where should the pulse be checked by health care professionals in children?**

The carotid artery should be used in children greater than 1 year of age.

☐ **If a pulse is present, how often should breaths be given?**

At a rate of 12-20 per minute.

☐ **If a pulse is present, when should EMS be activated?**

If the patient is breathing, immediately.

☐ **What is the purpose of chest compressions?**

To circulate oxygen-containing blood to the vital organs.

☐ **What are the two theories for the mechanism of blood flow during chest compressions?**

Thoracic pump theory and cardiac pump theory.

☐ **To achieve optimal compressions, how should the child be positioned?**

Supine on a hard, flat surface.

PEDIATRIC LIFE SUPPORT

☐ **Where should you perform chest compressions on infants?**

Just below the intermammary line.

☐ **How far should the infant's sternum be depressed during compressions?**

1/3 the depth of the chest.

☐ **What should the rate of compressions be during infant CPR?**

At least 100 times per minute.

☐ **What should the compression to ventilation ratio be?**

30:2 for lone rescuer, 15:2 for two resuers.

☐ **Children fall into what age-range for the purposes of BLS?**

Ages 1 year to adolescence.

☐ **T/F: Compressions in children should be performed with the fingers.**

False. Compressions should be performed with the heel of the hand.

☐ **What should the compression rate in children be?**

At least 100 times per minute.

☐ **What should the compression to ventilation ratio in children be?**

30:2 for lone rescuer, 15:2 for two rescuers.

☐ **How often should the infant and child be reassessed?**

After 2 minutes of CPR.

☐ **How is airway patency maintained during compressions?**

The hand not doing compressions is used to maintain the head tilt.

☐ **T/F: The majority of deaths from foreign body aspiration occur in children less than five years of age.**

True.

PEDIATRIC EMERGENCY MEDICINE BOARD REVIEW

☐ **When should attempts to clear the airway be considered?**

When foreign-body aspiration is witnessed or strongly suspected and when the airway remains obstructed during rescue breathing attempts.

☐ **T/F: Relief of airway obstruction should be attempted for all patients with foreign body aspiration.**

False. Only if signs of complete obstruction are observed.

☐ **For infants, what is the method by which complete foreign body obstruction is relieved?**

A combination of back blows and chest thrusts.

☐ **Why should the Heimlich maneuver not be performed on infants?**

There is a potential for liver injuries.

☐ **Why should blind finger sweeps be avoided?**

The foreign body may be pushed back into the airway.

☐ **How should back blows be delivered in an infant?**

Hold the infant face down, resting on forearm. The head should be lower than the trunk. Deliver back blows between scapulae with heel of hand.

☐ **T/F: Chest thrusts in a choking infant are delivered in the same manner as chest compressions.**

True.

☐ **How many back blows should be given?**

Five.

☐ **How long should the rescuer continue the series of back blows and chest thrusts?**

Until the object is expelled or the infant becomes unresponsive.

☐ **How are foreign bodies expelled in the conscious child?**

Abdominal thrusts while standing behind victim.

☐ **Why is it difficult to maintain neutral c-spine position in infants and young children?**

Prominent occiput predisposes the neck to slight flexion when the child is placed on a flat surface; this is why it is recommended to place a small towel under the shoulders to maintain the neck in neutral position.

PEDIATRIC LIFE SUPPORT

☐ **What is the dosing regimen for Lidocaine in RSI?**

1.5 mg/kg IV.

☐ **What is the dosing regimen for ketamine is RSI?**

1-4 mg/kg IV/IM.

☐ **What are the main negative side effects of ketamine?**

Hallucinations, emergence reaction, increasing intracranial pressure, laryngospasm, and increased intraocular pressure.

☐ **What is the purpose of administering Lidocaine in RSI?**

To reduce ICP.

☐ **What are the three main neuromuscular blocking agents used in RSI?**

Succinylcholine, vecuronium, rocuronium.

☐ **Of the above medications, which is depolarizing?**

Succinylcholine.

☐ **What is the dosing regimen for succinylcholine?**

1mg/kg in children, 2mg/kg in infants, IV/IM (IM dose double the IV dose).

☐ **What is the duration of succinylcholine?**

5-10 minutes.

☐ **What is the onset of action of succinylcholine?**

Very rapid, 30-60 seconds.

☐ **What is the dosing regimen for vecuronium?**

0.1-0.2mg/kg IV/IM.

☐ **What is the duration of vecuronium?**

20-30 minutes.

PEDIATRIC EMERGENCY MEDICINE BOARD REVIEW

☐ **What is the onset of action of vecuronium?**

2-3 minutes if given as a rapid push.

☐ **What is the dosing regimen of rocuronium?**

0.6-1.2 mg/kg.

☐ **What is the duration of rocuronium?**

30-60 minutes.

☐ **What is the onset of action of rocuronium?**

1-2 minutes.

☐ **Your patient has been placed on a mechanical ventilator in preparation for transport. How long after the initial ventilatory settings have been made should an arterial blood gas analysis be completed?**

After ten to fifteen minutes.

☐ **When using a volume-cycled ventilator, what is the recommended tidal volume that should be delivered to the patient?**

6-8 ml/kg.

☐ **When discussing mechanical ventilators, what does the acronym PEEP stand for?**

Positive End-Expiratory Pressure.

☐ **How often should the blood pressure of a hemodynamically unstable child be assessed?**

Blood pressure should be assessed every 5 minutes until stable and every 15 minutes thereafter. Attention to the patient's heart rate must also be closely monitored.

☐ **Explain the advantage of inserting a nasogastric tube.**

The tube prevents gastric distention caused by air and is particularly useful during positive-pressure ventilation.

☐ **You are caring for an infant that weighs 8 kg. What is the proper volume and rate of IV fluid administration for a patient smaller than 10kg?**

The proper fluid requirement in a patient smaller than 10kg is 4 ml/kg per hour (e.g., the maintenance rate for an 8 kg baby is 4 ml x 8 kg = 32 ml/hour).

PEDIATRIC LIFE SUPPORT

☐ You have been asked to set the initial mechanical ventilator settings for a patient who is intubated. After assessing the patient's airway you determine the tube is properly placed and the patient has normal lung compliance. What is the recommended peak inspiratory pressure and inspiratory time for this patient with normal lung compliance?

Effective ventilation may be achieved at peak inspiratory pressures of 20 to 30 cm water and an inspiratory time of 0.5 to 1.0 second.

☐ When mechanical ventilatory support is provided for the patient with normal lungs, a respiratory rate of _____ to _____ breaths per minute is typically required for infants and a rate of _____ to _____ breaths per minute for children.

Infants: 20 to 30 breaths per minute.
Children: 16 to 20 breaths per minute.

☐ Intubation bypasses glottic function and eliminates the physiologic positive end-expiratory pressure (PEEP) created during normal coughing, talking and crying. What should the PEEP setting be on the ventilator to maintain adequate functional residual capacity of the lungs?

To maintain adequate functional residual capacity, a PEEP of 2 to 4 cm water should be provided when mechanical ventilation is initiated.

☐ List two primary advantages of using noninvasive monitoring devices such as pulse oximeters, transcutaneous oxygen and carbon dioxide monitors, and exhaled carbon dioxide monitors.

These tools are very useful because they allow continuous assessment of oxygenation, ventilation, or both.

☐ Noninvasive monitoring devices such as those referenced in the above question may provide inaccurate data as a result of what types of medical conditions or equipment problems?

Hypothermia, poor peripheral perfusion, or endotracheal tube obstruction or displacement.

☐ You are the attending physician in the ER. While evaluating a six year-old male (GCS 6), you observe the child has significant central nervous system depression following a fall from the front porch of his home. In addition to BLS care, what procedure should be initiated until intracranial pressure can be evaluated more thoroughly?

Intubate and hyperventilate with the PaCO2 maintained at 30-35mm Hg until intracranial pressure can be evaluated more thoroughly. Also elevate the HOB to 30-35 degrees.

☐ T/F: Hypotension is a late sign of shock in infant and children, and requires rapid intervention.

True.

☐ If central venous access has been established, continuous or intermittent measurement of what type of pressure may help guide fluid administration and titration of vasoactive support?

Right heart filling.

PEDIATRIC EMERGENCY MEDICINE BOARD REVIEW

☐ **When should an orogastric or nasogastric tube be inserted into a patient, and what is its purpose?**

If bowel sounds are absent, abdominal distention is present, or the patient requires mechanical ventilation, an orogastric or nasogastric tube should be inserted to prevent or treat gastric distention.

☐ **Why is blind nasogastric tube placement contraindicated in the patient with serious facial trauma?**

Intracranial tube placement may result.

☐ **Which two agencies should be responsible for the establishment of well-defined protocols for specific clinical situations that arise during transport of an ill or injured child?**

The referring facility and the tertiary care unit.

☐ **Once started, how long must a physician continue to administer CPR?**

As necessary until the victim is transferred to the care of other properly trained personnel or until cardiac death (as defined by cardiovascular unresponsiveness to acceptable resuscitative techniques) is determined.

☐ **What is the most frequently used site for central venous cannulation and why?**

The femoral vein is the most frequently used central vein because it is large, anatomically reliable and can be cannulated without interrupting resuscitation.

☐ **T/F: If a central line will soon be available, delay administering medications and fluids until it is established.**

False. Establishment of a central line should not delay the administration of resuscitation medications and fluids.

☐ **What equipment is important for the safe administration of fluids and medications in infants and young children?**

Infusion pumps or volume limiting devices such as minidrip chambers (if pumps are not available) should be used to safely administer fluid boluses and medication infusions in infants and young children.

☐ **During CPR in young children what strategy should be followed for establishing vascular access?**

In general, intraosseous access should be established following three attempts at peripheral vascular access (approximately 90 seconds).

☐ **Is intraosseous access appropriate for children older than 6 years of age?**

Yes. Keep in mind that intraosseous access becomes progressively more difficult in older children. If intraosseous access must be attempted in older patients, drill devices are the best to use.

PEDIATRIC LIFE SUPPORT

☐ **Does a particular site insure more rapid delivery of medications to the central circulation during CPR?**

There is no significant difference between peripheral, intraosseous or central access with respect to onset of drug action or peak drug levels.

☐ **What should always be done immediately following a bolus infusion?**

It is important to follow medication delivery with a 5 cc fluid bolus to flush the drug to the central circulation.

☐ **What is the most important factor in determining route of intravenous access?**

Rapid access should be sought using the technique with which the medical personnel are most familiar.

☐ **What is the indication for ETT medication administration?**

If IV/IO access cannot be established within 3-5 minutes, the ETT is an important route for administration of medications.

☐ **What resuscitation medications can be infused through the ETT?**

Lidocaine, epinephrine, atropine, naloxone (L.E.A.N.).

☐ **What veins should be attempted for peripheral access during resuscitation?**

Greater saphenous (medial surface of the ankle) and the median cubital (antecubital fossa of elbow).

☐ **When should scalp veins be used?**

Scalp veins are of value when nonemergent fluids and medications are needed or in postresuscitation treatment as a temporizing measure until definitive access is achieved.

☐ **What are the most common causes of sclerosis and skin sloughing following IV cannulation?**

Necrotizing medications such as calcium and dextrose.

☐ **What are complications of intraosseous cannulation?**

Tibial fracture, compartment syndrome, osteomyelitis, and skin cellulitis.

☐ **What site is generally recommended for the placement of an intraosseous needle?**

The anteromedial tibial surface 2-3 cm. below the tibial tuberosity is the preferred site for placement in children.

PEDIATRIC EMERGENCY MEDICINE BOARD REVIEW

☐ **What are signs that intraosseous insertion is successful?**

The subjective sense of a decrease in resistance as the needle passes through the cortex and into the marrow, the needle is stable in the bone and remains upright without support, marrow can be aspirated through the needle into a syringe and infusion through the needle does not show evidence of infiltration of the soft tissue.

☐ **What should you do if you are unsuccessful on your first attempt at intraosseous cannulation?**

This same site cannot be used again. The procedure should be attempted on the other leg.

☐ **What are two advantages of central venous access in resuscitation?**

Central venous access allows rapid delivery of drugs to the central circulation and also allows monitoring of central venous pressures.

☐ **Are central venous lines, peripheral lines, or intraosseous lines more subject to complications?**

Central lines are clearly associated with more serious complications then peripheral or intraosseous lines. The specific complications are a function of the site of the central line.

☐ **What complications are associated with all central lines?**

Local and systemic infection, bleeding, phlebitis, thrombosis, catheter fragment and embolism.

☐ **What complications are associated with central lines in the neck?**

Pneumothorax, hydrothorax, and hemothorax.

☐ **What general techniques have been employed in achieving central venous access in children?**

Both through the needle and over the needle catheter placement has been used to establish central venous access in children.

☐ **Are complications more common in central lines in children or adults?**

They are more common in younger patients.

☐ **What are the hazards of through the needle catheter placement?**

The catheter can be sheared off by the sharp needle during placement of a through the needle catheter.

☐ **When does catheter shear usually occur when placing a through-the-needle catheter?**

This usually occurs when resistance is encountered as the catheter is advanced and the catheter has been withdrawn through the needle.

PEDIATRIC LIFE SUPPORT

☐ **What is the most commonly used technique for central venous access in children?**

The Seldinger technique, which uses a guide wire to establish catheter placement.

☐ **What is the most frequently used site for central venous access and why?**

The femoral vein is most frequently used because of its lower incidence of complications than central access in the neck, as well as the ease of starting a line far from the head and neck, which would interfere with CPR and airway management.

☐ **What are the advantages of cannulating the external jugular vein?**

The external jugular vein is a superficial vessel that is easily visible. When cannulated there is a large IV that allows rapid infusion of fluid and medications.

☐ **Where is the saphenous vein most easily palpated?**

Anterior to the medial malleolus.

☐ **What two methods are used for cannulating the saphenous vein?**

Blind stick or through a cut-down.

☐ **What are complications of arterial catheters?**

Infection, embolus, arterial thrombus, ischemia or damage to the effective limb.

☐ **What is the most common complication of radial artery catheterization in children?**

Radial artery occlusion.

☐ **What are acceptable sites for arterial catheterization?**

Radial, brachial, axillary, femoral, dorsalis pedis, and posterior tibial arteries.

☐ **Which two of the these are preferred?**

Radial and femoral artery.

☐ **What is the most significant physiological change undergone by the neonate?**

The transition from fetal to neonatal circulation.

☐ **What system must become instantly functional in this transition?**

The respiratory system, essentially nonfunctional in utero, must suddenly initiate and maintain oxygenation and ventilation.

PEDIATRIC EMERGENCY MEDICINE BOARD REVIEW

☐ **What is required to assist the majority of term newborns in making this transition?**

Maintenance of temperature, suctioning of the airway, and mild stimulation.

☐ **Of the small number of newborns that require further intervention, most respond to what?**

Administration of a high concentration of inspired oxygen and ventilation with a bag and mask.

☐ **What measure may be necessary if oxygen and bagging don't work?**

Chest compressions.

☐ **What is the least commonly required intervention in newborn resuscitation?**

Resuscitative medications.

☐ **What does the "inverted pyramid" illustrate?**

The relative frequencies and priorities of neonatal resuscitation.

☐ **What must follow each step in the pyramid to prevent unnecessary intervention and potential complications?**

Reassessment.

☐ **If there is a history of particulate meconium in the amniotic fluid, what should the resuscitation team be prepared to do?**

Suction the mouth and nose and consider suctioning of the trachea under direct visualization.

☐ **Which critical metabolic derangement can be exacerbated by hypothermia in the newborn?**

Acidosis.

☐ **How should the newborn be positioned?**

On his or her back or side with the neck in a neutral position.

☐ **Why should hyperextension of the neck be avoided?**

It may produce airway obstruction.

☐ **When placing the infant on his or her back, what can you do to insure neutral position?**

A rolled blanket or towel may be placed under the back and shoulders, thus elevating the torso 3/4 or 1 inch off the mattress to extend the neck slightly.

PEDIATRIC LIFE SUPPORT

☐ **How should the infant be positioned if copious secretions are present?**

On his or her side with the neck slightly extended to allow secretions to collect in the mouth rather than in the posterior pharynx.

☐ **When should the trachea be suctioned if meconium staining is observed?**

Immediately before other resuscitative steps are taken.

☐ **If meconium is absent but suctioning is required to ensure a patent airway, should the mouth or the nose be suctioned first?**

The mouth.

☐ **If mechanical suction is used, what negative pressure should not be exceeded?**

-100mm Hg (-136 cm H20).

☐ **What negative response can be produced by deep suctioning of the oropharynx?**

A vagal response that can cause bradycardia and/or apnea.

☐ **How long should you perform suctioning?**

No longer than 3-5 seconds per attempt.

☐ **What should you monitor during suctioning?**

The infant's heart rate.

☐ **What should you do between suctioning attempts?**

Time should be allowed between suction attempts for spontaneous ventilation or assisted ventilation with 100% oxygen.

☐ **What three forms of mild stimulation will be effective in stimulating breathing in most newborns?**

Drying, warming and suctioning.

☐ **What are two additional safe methods of stimulation?**

Slapping or flicking the soles of the feet and rubbing the infant's back.

PEDIATRIC EMERGENCY MEDICINE BOARD REVIEW

☐ **What should you do if spontaneous and effective respirations are not established after a brief period (5 to 10 seconds) of stimulation?**

Positive-pressure ventilation.

☐ **If respirations are deemed adequate on assessment, what should you check next?**

Heart rate.

☐ **If respirations are inadequate or gasping is present, what should you do?**

Begin positive-pressure ventilation immediately.

☐ **If respirations are shallow or slow, what should you do?**

A brief period of stimulation may be attempted while 100% oxygen is administered. The rate and depth of respirations should increase after a few seconds of stimulation and administration of oxygen.

☐ **"The mere presence of respirations does not guarantee adequate ventilation." Explain.**

Shallow respirations may primarily ventilate airway dead space and thus provide inadequate alveolar ventilation, resulting in hypoxemia, hypercarbia, and slowing of the heart rate.

☐ **What vital sign is considered a reliable indicator of the newborn's degree of distress?**

Heart rate.

☐ **What should you do if the heart rate is less than 100 beats per minute?**

Positive-pressure ventilation with 100% oxygen should be initiated immediately.

☐ **What three methods are used in evaluating heart rate?**

1. Palpation of the pulse at the base of the umbilical cord.
2. Palpation of the brachial or femoral pulse.
3. Auscultation of the apical heart sound with a stethoscope.

☐ **What should you do if the heart rate is less than 60 bpm and not increasing rapidly despite effective ventilation with 100% oxygen?**

Chest compressions should be initiated.

☐ **T/F: An infant may be cyanotic despite adequate ventilation and a heart rate greater than 100 bpm.**

True.

- **What should you do if central cyanosis is present in a newborn with spontaneous respirations and an adequate heart rate?**

  100% oxygen should be administered until the cause of the cyanosis can be determined.

- **What is acrocyanosis?**

  Peripheral cyanosis.

- **Should you administer oxygen to an infant with peripheral cyanosis?**

  No, it is a common condition in the first few minutes of life and is not indicative of hypoxemia.

- **What is the purpose of the Apgar scoring system?**

  It enables rapid evaluation of a newborn's condition at specific intervals after birth.

- **What are the five objective signs measured by the Apgar?**

  Appearance (color)
  Pulse (heart rate)
  Grimace (reflex irritability)
  Activity (muscle tone)
  Respirations

- **At what intervals should the Apgar be measured?**

  One and five minutes of age.

- **When should you take additional scores?**

  If the five-minute Apgar score is less than 7, additional scores are obtained every five minutes for a total of twenty minutes.

- **Can the Apgar score be used to determine the need for resuscitation?**

  No. If resuscitative efforts are required, they should be initiated promptly and should not be delayed while the Apgar score is obtained.

- **How does prematurity affect the Apgar score?**

  In preterm infants the Apgar score is more likely to be affected by gestational age than asphyxia.

- **When should you be concerned about administering 100% oxygen in newborn resuscitation?**

  Never. If oxygen is needed during the resuscitation of a newborn, 100% should be used without concern for its potential hazards.

PEDIATRIC EMERGENCY MEDICINE BOARD REVIEW

☐ **T/F: Only warmed and humidified oxygen should be administered.**

False. Ideally, oxygen should be warmed and humidified, but this may not be possible in an emergency situation.

☐ **What methods can be used to administer oxygen to a newborn?**

A head hood, facemask attached to a non-self-inflating ("anesthesia") bag, or by a simple mask held firmly to the infant's face with at least 5 Lpm oxygen flow.

☐ **What are three indications for positive-pressure ventilation?**

1. Apnea or gasping respirations. 2. Heart rate less than 100 bpm. 3. Persistent central cyanosis despite administration of 100% oxygen

☐ **How can you determine the appropriate tidal volume for bagging a neonate?**

Begin bag-valve-mask ventilations carefully and determine visually the volume and force required to produce adequate chest expansion.

☐ **For most newborns, what is the initial pressure required for lung inflation?**

30-40 cm H2O. Less pressure (20 cm H2O) is usually required for subsequent breaths.

☐ **What should you do if adequate chest expansion is not achieved with initial assisted ventilation?**

The head and the facemask should be repositioned.

☐ **What should you do if repositioning the head and facemask fail to produce effective chest expansion?**

Suction further and increase inflation pressure.

☐ **What should you do if you still cannot achieve effective chest expansion and improvement in color and heart rate?**

Intubate immediately.

☐ **When should you insert an orogastric tube?**

If BVM positive-pressure ventilation is required for more than approximately two minutes or if gastric distention develops.

☐ **What size orogastric tube should be used?**

8F or 10F.

PEDIATRIC LIFE SUPPORT

☐ **Many self-inflating ventilation bags have a pop-off valve. At what pressure is this valve usually preset?**

30-45 cm H2O.

☐ **What is the problem with the pop-off valve in neonatal resuscitation?**

The initial inflation of a newborn's lungs may require higher inspiratory pressures, and the valve may prevent effective inflation unless it is occluded. Such bags should therefore have a pop-off valve that is easily bypassed.

☐ **What is the ideal sized ventilation bag for neonatal resuscitation?**

450 – 750 ml.

☐ **What is the range of neonatal tidal volume?**

6-8 ml/kg.

☐ **Why are facemasks with cushioned rims recommended?**

They facilitate creation of an effective seal between face and mask.

☐ **Should tapered tubes be used for neonates?**

No, only tubes with a uniform internal diameter.

☐ **What is the preferred technique for performing chest compressions in the neonate and small infant?**

Two thumbs placed on the middle third of the sternum, with the fingers encircling the chest and supporting the back. The thumbs should be positioned side by side on the sternum just below the nipple line.

☐ **What anatomical feature is important to avoid when performing chest compressions, and why?**

The xiphoid portion of the sternum. May cause liver damage.

☐ **What is the depth of neonatal compression?**

One third the anterior-posterior diameter of the chest or to a depth that generates a palpable pulse.

☐ **What is the rate of neonatal compressions?**

120 times per minute.

☐ **T/F: The relaxation phase should be longer than the compression phase.**

False, they should both be smooth and equal.

☐ **T/F: The compressing thumbs or fingers should not be lifted off the sternum during the relaxation phase.**

True.

☐ **When should you cease compressions?**

When the heart rate reaches 60 or more bpm.

☐ **What is the most common cause of bradycardia and shock in the neonatal period?**

Profound hypoxemia.

☐ **When should medications be administered in cases of bradycardia and shock?**

If, despite adequate ventilation with 100% oxygen and chest compressions, the heart rate remains less than 60 bpm.

☐ **What is the preferred site for vascular access during neonatal resuscitation?**

The umbilical vein, because it is easily located and cannulated.

☐ **What preparation is required prior to umbilical catheterization?**

Skin prep, draped appropriately and the cord trimmed with a scalpel blade 1 cm above the skin attachment and held firmly with a ligature to prevent bleeding.

☐ **How do you identify the umbilical vein?**

It is a thin walled single vessel. In contrast, the umbilical arteries are paired, have thicker walls, and are often constricted. The lumen of the vein is larger than that of the arteries; thus the vessel that continues to bleed after the cord is cut is usually the vein. It is typically at the 12 o'clock location.

☐ **What size catheter should be used?**

A 3.5F or a 5F umbilical catheter.

☐ **How is the catheter prepared for insertion?**

It is flushed with heparinized saline (0.5 to 1 U/ml) and attached to a three-way stopcock.

☐ **How far should the catheter be inserted?**

So that the tip is just below the skin and blood can be readily aspirated and any air bubbles evacuated. The umbilical venous catheter is inserted only until a good blood return is obtained. This should correspond to a depth of insertion of 1 to 4 cm.

PEDIATRIC LIFE SUPPORT

☐ **What complication can result from inserting the catheter too far?**

Advancement of the catheter tip into the portal vein or hepatic circulation. Neonatal liver injury (including hemorrhage) has been linked with the administration of hypertonic and alkaline solution (e.g., THAM or sodium bicarbonate into the portal vein, so this position of the catheter tip is avoided).

☐ **What should be suspected if free blood return is absent?**

A wedged hepatic position.

☐ **What should you do if you suspect such wedging?**

You should withdraw the catheter to a position where blood can be freely aspirated.

☐ **How long can the umbilical catheter remain in place?**

It should be withdrawn as soon as possible after resuscitation to minimize the danger of infection or portal vein thrombosis.

☐ **What is the advantage of cannulating the umbilical artery?**

It enables monitoring of blood pressure and blood sampling for blood gas analysis and evaluation of acid-base balance.

☐ **What is the disadvantage of cannulating the umbilical artery?**

It is time-consuming and more difficult than venous cannulation.

☐ **What other sites of cannulation can be used?**

Peripheral veins in the extremities and scalp.

☐ **If fluid and medications are required but umbilical venous or arterial access cannot be obtained, what should you do?**

Attempt intraosseous cannula placement.

☐ **What route of medication administration should be used when vascular access cannot be achieved?**

Endotracheal and intraosseous.

☐ **What two drugs can be administered to the neonate by this route?**

Epinephrine and naloxone.

PEDIATRIC EMERGENCY MEDICINE BOARD REVIEW

☐ **What epinephrine dosing regimen should be used for the endotracheal route in neonates?**

The dose is 0.01 to 0.03 mg/kg (1:10,000) or 0.1 to 0.3 ml/kg IV every 3 to 5 minutes. The drug should be diluted to a volume of 3-5 ml of normal saline (use smallest volume for smallest babies), followed by several positive-pressure ventilations.

☐ **What are the indications for epinephrine in neonatal resuscitation?**

Asystole or spontaneous heart rate less than 60 bpm despite adequate ventilation with 100% oxygen and chest compressions.

☐ **What is the dose?**

0.01 to 0.03 mg/kg (0.1 to 0.3 ml/kg of the 1:10,000 solution). May be repeated every 3 to 5 minutes if required.

☐ **When is high dose epinephrine appropriate?**

Many randomized controlled studies failed to show improved outcomes routinely. However, may be considered for special circumstances suggesting catecholamine resistant condition eq. anaphylaxis. Known alpha or beta blocker overdose, severe sepsis already treated with high dose pressors.

☐ **Does the current data support the use of high dose epinephrine in neonatal resuscitation?**

Data is inadequate at present to evaluate the efficacy of high doses of epinephrine in newborns, and the safety of these doses has not been established.

☐ **What is the primary complication of high dose epinephrine in newborns?**

Prolonged hypertension, which may in turn lead to complications such as intracranial hemorrhage in preterm infants.

☐ **What are the two primary volume expanders for neonates?**

1. O-negative blood crossmatched with mother's blood. 2. Normal saline or Ringers lactate.

☐ **What is the dosage for volume expanders in neonatal resuscitation?**

10 ml/kg. Reassess and administer additional boluses as needed.

☐ **What are the routes of administration for naloxone?**

Intravenous, intraosseous, via the endotracheal tube, subcutaneous, or intramuscular (if perfusion is adequate).

☐ **What is the dosing of naloxone?**

0.1 mg/kg. The initial does may be repeated every 2 to 3 minutes as needed.

PEDIATRIC LIFE SUPPORT

☐ **In addition to epinephrine and naloxone, what other drugs are useful in neonatal resuscitation?**

Glucose (Hypoglycemia should always be considered in a neonate. There is no evidence that atropine, calcium, or sodium bicarbonate is beneficial in the acute phase of neonatal resuscitation at delivery.

☐ **What percentage of all deliveries are complicated by the presence of meconium in the amniotic fluid?**

Approximately 12%.

☐ **Meconium is found in the trachea of what percent of newborns with meconium staining despite suctioning?**

20-30%.

☐ **T/F: Suction should be applied directly to the endotracheal tube.**

True.

☐ **How is this accomplished?**

With a meconium aspirator.

☐ **What should you do if there is a significant amount of meconium?**

Repeat intubation and suction while withdrawing the ET tube, until the aspirated material is clear or until the heart rate indicates that resuscitation should proceed without delay.

☐ **What should you do if the patient's condition rapidly deteriorates before all meconium is removed?**

When an infant's condition is unstable, it may not be possible to clear the trachea of all meconium before positive-pressure ventilation must be initiated.

☐ **Why should an orogastric tube be placed in these infants?**

To empty the stomach, since it may contain meconium that could later be regurgitated and aspirated.

☐ **Why are preterm infants more subject to intracranial bleeds?**

Because the brain of the preterm infant has a fragile subependymal germinal matrix that is vulnerable to bleeding when injured by hypoxemia, rapid changes in blood pressure, or wide fluctuations in serum osmolality.

☐ **Why are preterm infants more likely to become hypothermic?**

Because their ratio of body surface area to volume is higher than that of full-term neonates.

PEDIATRIC EMERGENCY MEDICINE BOARD REVIEW

☐ **Because of this increased risk of intracranial bleeding, what should be avoided in these patients?**

Administration of hyperosmolar solutions or large boluses of volume expanders.

☐ **What are the four most common complications in the postresuscitation period?**

(DOPE pneumonic)
1. Dislodgment: Endotracheal tube migration
2. Obstruction: Tube occlusion by mucus or meconium
3. Pneumothorax
4. Equipment: Oxygen not being delivered or other equipment problems

☐ **Why is a pneumothorax difficult to diagnose by auscultation?**

Because breath sounds in the newborn are transmitted from all areas of the lung through the thin chest wall.

☐ **When should you suspect pneumothorax?**

If a newborn deteriorates after an initial good response to ventilation or fails to respond to resuscitative efforts. Additional signs of pneumothorax include unilateral decrease in chest expansion, altered intensity or pitch of breath sounds, and increased resistance to hand ventilation.

☐ **Your patient has been placed on a mechanical ventilator in preparation for transport. How long after the initial ventilatory settings have been made should an arterial blood gas analysis be completed?**

After fifteen minutes.

☐ **What is the purpose of administering Lidocaine in RSI?**

Reduce ICP.

☐ **When using a volume-cycled ventilator, what is the recommended tidal volume that should be delivered to the patient?**

6-8 ml/kg.

☐ **List two primary advantages of using noninvasive monitoring devices such as pulse oximeters, transcutaneous oxygen and carbon dioxide monitors, and exhaled carbon dioxide monitors.**

These tools are very useful because they allow continuous assessment of oxygenation, ventilation, or both.

☐ **Noninvasive monitoring devices such as those referenced in the above question may provide inaccurate data as a result of what types of medical conditions or equipment problems?**

Inaccurate results may occur as a result of hypothermia, poor peripheral perfusion, or endotracheal tube obstruction or displacement.

PEDIATRIC LIFE SUPPORT

☐ **Urine output often correlates with the effectiveness of renal and systemic perfusion. What is the preferred method of monitoring urine output?**

An indwelling catheter.

☐ **What is the most common cause of shock in children?**

Hypovolemic shock.

☐ **What kind of shock is caused by peripheral vasodilatation resulting in venous pooling of blood and a decrease of blood returning to the central circulation?**

Neurogenic shock.

☐ **What types of shock result from both vasodilatation and increased capillary permeability causing plasma losses from the vascular space and into the interstitium caused?**

Septic shock and anaphylactic shock.

☐ **What is shock resulting from inadequate heart (pump) function called?**

Cardiogenic shock.

☐ **T/F: Blood products are considered the first choice for the management of hypovolemia.**

False. Crystalloids should be tried first.

☐ **When is blood considered the ideal fluid replacement in volume loss?**

When trauma victims in hypovolemic shock do not respond to crystalloid management or trauma patients present in decompensated shock.

☐ **What is the blood type that may be administered without crossmatch?**

O-negative.

☐ **Rapid infusion of cold blood or blood products containing citrate in large volumes may result in what two major complications?**

Hypothermia and hypocalcemia.

☐ **When is volume therapy indicated?**

When the child demonstrates signs and symptoms of shock.

☐ **What are five significant signs of hypovolemic shock in a child?**

Tachycardia, pale, mottled, cool skin, delayed capillary refill, diminished peripheral pulses, and altered mental status.

☐ **T/F: Optimum vascular access in a child requires only one large bore peripheral line.**

False. At least two are required.

☐ **What is the fluid bolus dose of crystalloid for the management of the symptomatic hypovolemic child?**

20ml/kg IV in less than 20 minutes.

☐ **How many times may fluid boluses of crystalloid be repeated during the first hour of trauma to manage volume losses in a hypovolemic child before giving blood?**

Three times.

☐ **T/F: A child in septic shock may require 60 to 80 ml/kg during the first hour of resuscitation.**

True.

☐ **What should you do following each volume bolus?**

Reassess perfusion status of the child. Evaluate for effectiveness of therapy.

☐ **T/F: Large volumes of dextrose containing solutions are particularly useful during volume resuscitation.**

False. They can be harmful because of their hypertonic effects.

☐ **What would be the appropriate management of a patient with a respiratory acidosis?**

Adequate oxygenation and appropriate ventilation.

☐ **What would be the appropriate management of a patient with a metabolic acidosis secondary to poor perfusion?**

Correct the perfusion problem.

☐ **What are the indications for the administration of epinephrine?**

Cardiac arrest (Alpha), symptomatic bradycardia that will not respond to oxygenation and ventilation, and hypotension not responding to volume resuscitation.

PEDIATRIC LIFE SUPPORT

- **How do you determine a base deficit in the setting of a metabolic acidemia?**

    Calculate the difference between the predicted pH (7.40) and the measured pH.
    Multiply the difference by 67 (constant).
    This will give you the patient's base deficit.
    Example: Measured pH 7.18
    Predicted pH 7.40
    Difference: -0.22
    -0.22 x 67 = -14.7 Base Deficit

- **How often should epinephrine be administered in a cardiac arrest resuscitation?**

    Every 3 - 5 minutes.

- **What are two post epinephrine administration side effects?**

    Hypertension and tachycardia.

- **What narcotic agent is recommended in kids?**

    Fentanyl citrate (Sublimaze) 2-4 mcg/kg IV or IM.

- **What is the duration of action of Fentanyl?**

    30 minutes to 1 hour.

- **What is the advantage of Fentanyl over other opioids?**

    This medication does not cause hypotension in low or high doses.

- **What is the dosing regimen for Midazolam?**

    0.1-0.2mg/kg (maximum 4mg) IV or IM, 1-2 hours.

- **What is the most significant side effect of Midazolam?**

    Respiratory depression.

- **T/F: Midazolam has analgesic as well as sedative properties.**

    False. That is why it is important to always give an analgesic in addition when performing painful procedures.

- **Why is it important to be able to adequately ventilate a patient who has received sodium bicarbonate?**

    NaHCO3 (Sodium Bicarbonate) generates an increase in CO2 production as the hydrogen ion is buffered – adequate ventilation is necessary to remove the additional load of CO2.

PEDIATRIC EMERGENCY MEDICINE BOARD REVIEW

☐ **When should sodium bicarbonate be considered for administration?**

When severe acidosis is associated with prolonged cardiac arrest, shock, hyperkalemia or tricyclic antidepressant toxicity.

☐ **How fast should sodium bicarbonate be administered?**

Slowly over 1-2 minutes.

☐ **T/F: Excessive administration of sodium bicarbonate may result in metabolic alkalosis.**

True.

☐ **T/F: Administration of sodium bicarbonate can result in lowering serum potassium.**

True.

☐ **What are two cardiovascular conditions for which atropine administration is indicated?**

Symptomatic second degree heart blocks and hemodynamically significant bradycardia.

☐ **What is the benefit of atropine administration to a child during endotracheal intubation attempts?**

Atropine can prevent vagally mediated bradycardia.

☐ **What is the recommended dose of atropine?**

0.02mg/kg IV or intraosseous (IO).

☐ **What is the minimum IV/IO single dose of atropine for a child?**

0.1mg.

☐ **What is the maximum total dose of atropine for a child?**

0.04 mg/kg.

☐ **How often may atropine be repeated?**

3 - 5 minutes after initial administration if symptoms persist.

☐ **What may atropine administered at lower than recommended doses do to the heart rate?**

Cause a paradoxical slowing of the heart rate.

PEDIATRIC LIFE SUPPORT

☐ **T/F: The pupillary dilatation associated with atropine will not react (constrict) to light reflex.**

False. The pupils will still constrict.

☐ **Naloxone (Narcan) is indicated for what condition?**

Narcotic (opioid) toxicity induced symptoms.

☐ **What are three significant symptoms associated with narcotic (opioid) intoxication?**

Respiratory depression, CNS depression, and hypoperfusion.

☐ **What is the onset of effect for naloxone (Narcan)?**

< 2 minutes.

☐ **What is the duration of activity for naloxone (Narcan)?**

45 – 120 minutes.

☐ **By what routes may naloxone (Narcan) be administered?**

IV, IO and ETT.

☐ **What is the recommended IV/IO dose of naloxone (Narcan) for infants and children up to 20kg?**

0.1mg/kg.

☐ **What is the recommended infusion rate for Narcan?**

0.04 - 0.16 mg/kg/hour.

☐ **What may occur after administration of naloxone (Narcan) if the narcotic effect is abruptly reversed?**

Acute narcotic withdrawal.

☐ **What are the symptoms of acute narcotic withdrawal?**

Nausea and vomiting, tachycardia, hypertension, seizures and cardiac dysrhythmias.

☐ **Why is hypoglycemia bad?**

Because it is important for cells to function normally especially in the brain. Hypoglycemia can precipitate seizure activity and depress myocardial function.

PEDIATRIC EMERGENCY MEDICINE BOARD REVIEW

☐ **When should glucose administration be considered?**

When hypoglycemia is present or the infant or child fails to respond to standard resuscitation measures.

☐ **What is the dosage range of glucose?**

0.5 - 1.0 gm/kg IV or IO.

☐ **What is the maximum recommended concentration of glucose for administration to children?**

25% (D25W) for children, D10W for neonates and infants, D50W for adolescents.

☐ **What is the dilution to reduce the concentration to 10% (D10W)?**

1:4 with sterile water.

☐ **What are four conditions that can lead to poor outcomes if high concentration glucose (D50/D25) is administered to children?**

Children with head injuries, near drowning (submersion), stroke, and shock.

☐ **What are four indications for administration of calcium?**

Documented or suspected hypocalcemia, hyperkalemia, hypermagnesemia and calcium channel blocker overdose.

☐ **10% calcium chloride is equal to how many milligrams per milliliter?**

100 mg/ml.

☐ **What is the recommended dose of calcium chloride?**

20 mg/kg IV.

☐ **How fast should calcium chloride be infused?**

Do not exceed 100 mg/min.

☐ **What may happen if calcium is administered too fast?**

Bradycardia or asystole may occur.

☐ **Why should calcium only be administered through a large, well-secured intravenous line?**

Calcium can cause significant chemical damage if it infiltrates into surrounding tissue.

PEDIATRIC LIFE SUPPORT

☐ **What does prostaglandin E1 do?**

Prevents closure of the ductus arteriosus in newborns.

☐ **What are the indications for administration of prostaglandin E1?**

Infants with congenital cardiovascular disease with ductal dependent lesions.

☐ **What are the signs and symptoms, in the newborn, associated with congenital cardiovascular abnormalities that could indicate the need for administration of prostaglandin E1?**

Cyanosis or shock.

☐ **How should prostaglandin E1 be administered?**

Continuous intravenous infusion.

☐ **Why should prostaglandin E1 be administered by continuous intravenous infusion?**

It has a very short half-life.

☐ **What is the effective dose range of prostaglandin E1?**

0.05 - 0.10 mcg/kg/min.

☐ **Because prostaglandin E1 may cause apnea, what should you be prepared to do during its administration?**

Secure the airway with an endotracheal tube and oxygenate/ventilate the infant.

☐ **What are five potential adverse reactions associated with epinephrine infusion?**

Tachycardia, ventricular dysrhythmias, profound peripheral vasoconstriction, compromised renal and hepatic blood flow, and local infiltration may cause ischemic tissue damage.

☐ **What are three indications for the administration of lidocaine by IV bolus?**

Ventricular dysrhythmias, ventricular tachycardia, and ventricular fibrillation (after defibrillation).

☐ **What is the recommended dose of lidocaine IV bolus?**

1mg/kg (not to exceed 3mg/kg).

☐ **When should lidocaine infusion be started?**

Following effective lidocaine bolus administration.

☐ **What are three potential adverse reactions associated with lidocaine administration?**

Myocardial depression, hypotension, and central nervous system manifestations, such as drowsiness, disorientation, muscle twitching and seizures.

☐ **What are the indications for the administration of adenosine?**

Supraventricular tachycardia (heart rates >220 BPM) with or without evidence of poor perfusion.

☐ **What is the recommended dose of adenosine?**

0.1 - 0.2mg/kg IV rapid bolus.

☐ **What are the indications for the use of Amiodarone in children?**

Wide range of atrial and ventricular arrhythmias, particularly ectopic atrial tachycardia, junctional ectopic tachycardia, and ventricular tachycardia.

☐ **What are the two main precautions when using Amiodarone?**

Hypotension and a prolonged QT interval.

☐ **What is the dosing of Amiodarone in refractory pulseless VT, VF?**

5mg/kg rapid IV/IO bolus.

☐ **What is the dosing of Amiodarone for perfusing supraventricular and ventricular arrhythmias?**

Loading dose: 5mg/kg IV/IO over 20 to 60 minutes (repeat to a maximum of 15mg/kg per day IV).

☐ **List the six ways the pediatric airway differs from the adult airway.**

1. The airway is smaller.
2. The tongue is relatively larger.
3. The larynx is more cephalad in position.
4. The epiglottis is short, narrow and angled away from the trachea.
5. The vocal cords attach lower anteriorly.
6. In children less than 10 the narrowest portion of the airway is subglottic.

☐ **Why should a straight laryngoscope blade be used in children?**

The high position of the larynx makes the angle between the tongue and glottis more acute. This has been called into question regarding all age groups but the neonates.

☐ **Should endotracheal tube size be based on the size of the glottic opening or the size of the cricoid ring?**

The cricoid ring, as this is the narrowest part of the trachea in children.

PEDIATRIC LIFE SUPPORT

☐ **How is airflow resistance related to airway radius?**

Inversely proportional to the fourth power of the airway radius during laminar flow (quiet breathing), and inversely proportional to the fifth power of airway radius during turbulent airflow.

☐ **What happens to functional residual capacity when respiratory effort is diminished?**

It is reduced.

☐ **Why is the pediatric airway susceptible to dynamic collapse during airway obstruction?**

High compliance of the airway.

☐ **Why do children have a high oxygen demand per kilogram?**

Their metabolic rate is high.

☐ **Does hypoxemia occur more quickly in the child or adult in response to apnea?**

The child, secondary to high oxygen consumption.

☐ **Why should children in respiratory distress be allowed to maintain a position of comfort?**

This position usually allows the maximal airway patency and minimizes respiratory effort.

☐ **What can occur if breaths with a bag-valve-mask are not coordinated with the child's breathing efforts?**

Coughing, vomiting, laryngospasm, and gastric distension.

☐ **Why is pulse oximetry an important technique for monitoring respiratory insufficiency?**

It provides continuous evaluation of arterial oxygen saturation.

☐ **T/F: Pulse oximeters reflect the effectiveness of ventilation.**

False, they only evaluate level of oxygenation.

☐ **When is the pulse oximeter's accuracy limited?**

In the presence of methemoglobinemia, carbon monoxide poisoning or in shock states.

☐ **Why is the pulse oximeter not useful during shock or poor perfusion?**

It requires pulsatile blood flow to determine oxygen saturation.

PEDIATRIC EMERGENCY MEDICINE BOARD REVIEW

☐ **Where may the oximeter probe be placed if a signal is not detected at the fingertip?**

Ear lobe, nares, cheek, tongue and foot.

☐ **What is the most accurate method of determining arterial oxygen concentration?**

Arterial blood gas.

☐ **Why is a low flow oxygen delivery system insufficient to meet all inspiratory flow requirements?**

Because with low flow systems room air is entrained and mixed with the oxygen decreasing the concentration of oxygen to be inhaled by the patient.

☐ **What is the maximum inspired oxygen concentration achieved with a nonrebreather mask?**

95%.

☐ **What is the indication for use of an oral airway?**

In an unconscious child with airway obstruction after usual airway maneuvers have failed.

☐ **Why should an oral airway not be used in a conscious child?**

May stimulate gagging or vomiting.

☐ **How is the appropriately sized oral airway chosen?**

Measure from the level of the central incisors, the bite block segment is parallel to the hard palate and the curved portion should reach the angle of the jaw.

☐ **What is the appropriate length for a nasal airway?**

Distance from the tip of the nose to the tragus of the ear.

☐ **Why should the space under the mask used for ventilation be small?**

To decrease dead space and minimize rebreathing of exhaled gases.

☐ **From where to where should the appropriately sized ventilation mask extend?**

The bridge of the nose to the cleft of the chin.

☐ **When performing bag-valve-mask ventilation, why should pressure on the submental area be avoided?**

Can cause the tongue to be pushed into the posterior pharynx leading to obstruction or airway compression can occur.

PEDIATRIC LIFE SUPPORT

☐ **What patient position is appropriate during BVM of infants and toddlers?**

Supine and head and neck in the neutral sniffing position.

☐ **What should be done if effective BVM ventilation cannot be achieved?**

Reposition head, make sure mask is snug, lift jaw, consider suctioning, ensure bag is functioning and connected to gas source.

☐ **How can gastric inflation be minimized in the unconscious child?**

Use appropriate sized equipment, only inflate the lungs with a volume necessary to initiate chest rise. Ventilate at an age appropriate rate, apply cricoid pressure.

☐ **What can sudden decreases in lung compliance indicate?**

Right main bronchus intubation, obstructed endotracheal tube, and pneumothorax.

☐ **What are some of the advantages of ventilation via an endotracheal tube?**

The airway is isolated, potential for aspiration is reduced, ventilations and chest compressions can be interposed efficiently, inspiratory time and peek pressures can be controlled and PEEP can be delivered.

☐ **What are six indications for intubation?**

1. Inadequate CNS control of ventilation.
2. Loss of protective airway reflexes.
3. Respiratory failure.
4. Need for high ventilatory pressures or PEEP.
5. Need for mechanical ventilatory support.
6. Allows for stabilization of an airway that has the potential to deteriorate in transport or during special procedures at a hospital.

☐ **When should a cuffed endotracheal tube be used?**

In all age groups other than neonates.

☐ **At what pressure should an air leak occur in the intubated patient?**

20 to 30 cm H2O.

☐ **What does the absence of an air leak indicate?**

The cuff is inflated excessively, the ET tube is too large, or laryngospasm is occurring around the tube.

PEDIATRIC EMERGENCY MEDICINE BOARD REVIEW

☐ **If visual inspection is used for choosing the endotracheal tube size, what part of the child's anatomy should be chosen for approximating the appropriate diameter?**

The width of a child's little fingernail.

☐ **What formula can be used to choose ET tube size in children?**

(Age [years]/4)+4 (Uncuffed).
(Age/4) + 3 (Cuffed).

☐ **What is the most accurate way of determining ET tube size?**

Using a Broselow tape for weight estimate and then reading the appropriate tube size for that weight.

☐ **How can the ET tube size be used to estimate the depth of insertion of the tube?**

Internal diameter x 3 = depth of insertion (cm).

☐ **What should be done if bradycardia occurs during an intubation attempt?**

The procedure should be interrupted and the patient should be ventilated with 100% oxygen via bag-valve-mask.

☐ **When should an intubation attempt be interrupted?**

If the patient develops hypoxemia, cyanosis, pallor, or a decreased heart rate.

☐ **To directly visualize the glottis, what three structures must be aligned?**

The axes of the mouth, pharynx and trachea.

☐ **How should children less than two years old without trauma be placed for intubation?**

On a flat surface with the chin in sniffing position.

☐ **Why should the blade not be inserted into the esophagus and withdrawn to visualize the epiglottis?**

This practice increases the risk of laryngeal trauma.

☐ **Where should the tip of the curved blade (Macintosh) be placed?**

In the vallecula.

☐ **What is done with the tip of the straight blade (Miller)?**

Used to lift the epiglottis to visualize the glottic opening.

PEDIATRIC LIFE SUPPORT

☐ **From which direction should the endotracheal tube be inserted?**

From the right corner of the mouth.

☐ **Where is the black glottic marker on the endotracheal tube placed?**

At the level of or just below the vocal cords.

☐ **How is the position of the endotracheal tube assessed after intubation?**

Observation of symmetrical chest movement, auscultation of equal breath sounds, documentation of absent gurgling sounds over the stomach, and measurement of end-tidal carbon dioxide level.

☐ **When can endotracheal carbon dioxide levels be low?**

In low cardiac output states.

☐ **What problems should be considered when intubation is confirmed but oxygenation or ventilation is inadequate?**

The endotracheal tube is too small, the pop off valve on the resuscitation bag is not depressed, A leak is present in the connections of the bag-valve device, inadequate tidal volume is provided by the operator, and lack of lung expansion or lung collapse is occurring for other reasons.

☐ **Why shouldn't oxygen powered breathing devices be used in children?**

High airway pressures may produce gastric distension or tension pneumothorax.

☐ **How can displacement of the endotracheal tube be detected?**

By noting the distance marker number at the lips, deterioration in patient status, by use of end tidal CO2 device, by decreased oxygen saturation, by clinical assessment that the tube is in the right mainstem (decreased breath sounds on the right versus the left) or by noting gurgling sounds in the stomach indicating esophageal displacement of the tube.

☐ **When should a tension pneumothorax be suspected?**

When any intubated patient deteriorates suddenly during positive pressure ventilation.

☐ **What are the clinical signs of tension pneumothorax?**

Severe respiratory distress, hyperresonance to percussion, diminished breath sounds on the affected side, JVD and deviation of the trachea and mediastinum away from the affected side.

☐ **What is the treatment for tension pneumothorax?**

Immediate needle decompression.

PEDIATRIC EMERGENCY MEDICINE BOARD REVIEW

☐ **For what clinical conditions might cricothyroidotomy be effective?**

Airway obstruction caused by foreign body, severe orofacial injuries, infection and laryngeal fracture.

☐ **What size needle should initially be placed through the cricothyroid membrane?**

20 gauge.

☐ **What size cannula should eventually be placed?**

At least a 14 gauge.

☐ **How is the cannula connected to a bag valve device?**

Using a 3 mm endotracheal tube adapter.

☐ **What gas flow rate should be used for oxygenation in needle cricothyroidotomy?**

1 to 5 L/min (100cc/kg).

☐ **If a patient on a ventilator experiences a problem, what should be done?**

Patient should be manually ventilated with an ambu bag using 100% oxygen.

☐ **If the endotracheal tube is occluded and patency can't be restored, what should be done?**

Remove tube and ventilate with bag-valve-mask until reintubation.

☐ **What equipment should be kept at the bedside of a patient with a tracheostomy?**

A pair of scissors and a new tracheostomy tube.

☐ **If recannulation of a patient's tracheostomy stoma cannot be accomplished, what should be done?**

Try a smaller size stoma first; if this doesn't work orally intubate, or perform bag-valve-mask ventilation with gauze occluding the stoma. If there is no chest rise try placing a small mask over the stoma and performing mask to stoma ventilation using a bag-valve device.

☐ **Why is a cricothryoidotomy contraindicated in a child < 10 years of age?**

High incidence of subglottic stenosis.

☐ **Decompensated shock is characterized by what conditions?**

Hypotension and low cardiac output.

PEDIATRIC LIFE SUPPORT

☐ **In an unconscious, non-breathing patient with suspected cervical injury, which maneuver should be used to open the airway?**

Jaw thrust with cervical spine immobilization.

☐ **Why shouldn't blind finger sweeps of the mouth be performed for manual removal of a foreign body?**

Because the foreign body may be pushed back into the airway, resulting in further obstruction.

☐ **When should atropine be used to treat bradycardia?**

Only after adequate ventilation and oxygenation have been established, since hypoxemia is the most common cause of bradycardia in children.

☐ **Name some complications associated with intraosseous cannulation and infusion.**

Tibia fracture, compartment syndrome, skin necrosis and osteomyelitis. This has been reported in less than 1% of patients.

☐ **What are the most common causes for deterioration of ventilatory status in a stable, intubated patient?**

Endotracheal tube dislodgment, occlusion, pneumothorax and equipment failure.

☐ **Hypoxemia with a normal alveolar-arterial oxygen difference is the result of what?**

Hypoventilation.

☐ **What is the narrowest part of the respiratory tract in children?**

The inferior ring portion of the cricoid cartilage.

☐ **In what phase of respiration is stridor observed.**

Inspiration.

☐ **What causes stridor?**

With extrathoracic airway obstruction, the pressure inside the extrathoracic part of the airway is negative relative to the level of atmospheric pressure. This results in further narrowing of the larynx during inspiration and therefore, stridor.

☐ **A neonate has Apgar scores of 8 and 8. He is actively crying, but cyanosis and retractions appear when he is quiet. What is the most likely diagnosis?**

Choanal atresia.

PEDIATRIC EMERGENCY MEDICINE BOARD REVIEW

☐ **Which anomalies of the great vessels result in airway obstruction?**

1. Right aortic arch with or without a left ligamentum arteriosum.
2. Double aortic arch.
3. Anomalous innominate or left carotid artery.
4. Aberrant right subclavian.
5. Anomalous left pulmonary artery.

☐ **To achieve more than 90% FIO2 in a non-intubated patient, which oxygen device should be used?**

Non-rebreather mask with a tight fitting mask.

☐ **Name some skin manifestations of shock/hypoxemia.**

Circumoral pallor/cyanosis, mottled skin, distal cyanosis, prolonged capillary refill, diminished distal pulses.

☐ **A three-year-old presents with signs/symptoms of epiglottitis. Should intravenous access be a priority?**

No. Intravenous access is an absolute contraindication in suspected epiglottitis. The procedure is to put the patient in a comfortable position with supplemental oxygen and minimize the level of anxiety in the patient. All invasive procedures are to be done in the operating room.

☐ **In a child who is hypoglycemic and hypotensive, is it appropriate to use a solution containing dextrose for fluid resuscitation?**

No. The use of a glucose containing solution for fluid resuscitation will result in an osmotic diuresis secondary to hyperglycemia.

☐ **What is the appropriate treatment for a hypoglycemic child?**

D50W, 1.0 ml/kg through a central line, or D25W, 2.0 ml/kg through a central or peripheral line, use D10 in infants.

☐ **What are the relative contraindications to the use of a nasopharyngeal airway?**

Suspected airway trauma, adenoidal hypertrophy, and bleeding diatheses.

☐ **After intubating a 1 month old baby for respiratory failure, you confirm appropriate tube placement by auscultation. However, CXR reveals the tube is the right mainstem bronchus. Why were you able to hear breath sounds bilaterally?**

In small infants, it is not uncommon to hear transmitted breath sounds. In the patients, the physician is advised not only to auscultate, but also to visualize adequate chest rise bilaterally.

PEDIATRIC LIFE SUPPORT

☐ **What is the amount of FiO2 each device can deliver?**

Nasal cannula: 30-40%.
Oxygen hoot/tent: up to 80-90%.
Simple mask: up to 40%.
Partial rebreathing mask: up to 60%.
Nonrebreathing mask: 65-95%.

☐ **What comprises the assessment of the circulatory system in the pediatric patient?**

Blood pressure, heart rate, peripheral and central pulses, capillary refill, and skin color.

☐ **What is the time-frame for a normal capillary refill?**

Less than three seconds.

☐ **What are some major complications of endotracheal intubation?**

Esophageal intubation/perforation, trauma to the teeth, emesiswith aspiration, and bradycardia.

☐ **Is there a difference between calcium chloride and calcium gluconate, in an emergency situation?**

Yes. The liver must first metabolize calcium gluconate before elemental calcium is available. Therefore, calcium chloride is the preferred drug during an emergency.

☐ **What is the dose of calcium chloride?**

0.2 – 0.3ml/kg (5 - 10ml) of a 10% solution (100mg/ml). If a peripheral route is used, the drug should be given slowly. Extravasation of calcium can lead to tissue damage and necrosis.

☐ **When inserting an intraosseous needle into the tibia, in which direction should the needle be placed?**

Caudal, away from the epiphysial plate. This will avoid damage to the growth plate of the tibia.

☐ **When is isoproterenol indicated?**

Bradyarrhythmias and heart block. It can also be used in small infants with heart failure, because of its chronotropic properties.

☐ **You are asked to assist in a precipitous delivery. The newborn is lethargic and unresponsive to tactile stimulation. The mother admits to recent heroin use. How should you proceed?**

The infant is most likely suffering from narcotic induced respiratory depression. Treatment is naloxone given either intravenously, intratracheally, intraosseously, subcutaneously, or intramuscularly, at a dose of 0.1 mg/kg.

PEDIATRIC EMERGENCY MEDICINE BOARD REVIEW

☐ **While hand-bagging a 14-year-old patient, you notice the stomach is getting distended. What steps should you take to remedy the situation?**

Apply gentle cricoid pressure to occlude the esophagus. Insertion of nasogastric or orogastric tube can also be used to decompress the stomach.

☐ **In the previous patient, why not just push on the epigastric area to deflate the stomach?**

Pressure applied to the stomach for decompression can lead to emesis and aspiration.

☐ **What is the normal systolic pressure in a neonate?**

Greater than 60 mm Hg.

☐ **Which actions are appropriate to provide tactile stimulation on a newborn in order to encourage the patient to cry?**

Flick or slap the sole of foot, or rub the back (preferably with a towel while drying the baby).

☐ **What three signs suggest that a newborn is responding to resuscitation?**

Increasing heart rate, spontaneous respirations, improving color.

☐ **Failure to respond to cardioversion is due to what?**

Hypothermia, hypoglycemia, acidosis, or hypoxia.

☐ **If a newborn fails to respond to epinephrine, what should be the next step?**

Repeat epinephrine, volume expansion, and sodium bicarbonate. Unlike older pediatric patients where bicarbonate is indicated only in documented metabolic acidosis, the use of bicarbonate is indicated when the patient is not responding to epinephrine, even if there is no documented acidosis.

☐ **Which sedatives are recommended for routine intubation?**

Valium 0.1 mg/kg, midazolam 0.1 mg/kg, lorazepam 0.1 mg/kg, morphine 0.1 mg/kg, fentanyl 1 mcg/kg.

☐ **A 4-year-old is brought to the emergency room with suspected foreign body aspiration. The patient is alert, cyanotic, and gasping for air. What should be the immediate treatment?**

Provide oxygen. As long as the patient is awake medical personnel should not intervene. Patient should be encouraged to continue coughing and breathing.

☐ **After a few minutes, the patient described in the previous question becomes unconscious. What should be the next step?**

Administer 5 back blows, followed by 5 abdominal thrusts until object is dislodged. When treating infants, administer 5 back blows, followed by 5 chest thrusts.

PEDIATRIC LIFE SUPPORT

☐ **Despite repeated efforts, the object has still not been dislodged. What should be the next procedure?**

An emergency needle cricothyrotomy or tracheostomy.

☐ **Which medications cannot be given intraosseously?**

Phenytoin and chloramphenicol.

☐ **Which site is used to evacuate a pneumothorax by using needle thoracentesis?**

The second intercostal space at the midclavicular line, or the fourth intercostal space at the anterior or midaxillary line.

☐ **Which drug overdoses are associated with cardiac arrhythmias?**

Cocaine, heroin, tricyclic antidepressants, digitalis, calcium channel-blockers, and beta-blockers.

☐ **Which laboratory studies can be obtained through an intraosseous needle?**

Electrolytes and blood cultures.

☐ **Are there any differences in the "ABC" approach to cardiopulmonary arrest in children (as compared to adults)?**

The airway and breathing are stressed more in children than is circulation. Except in the child with congenital cardiovascular anomalies, there are few purely cardiac causes of cardiopulmonary arrest.

☐ **What is the normal IV maintenance amount for a 25 kg child?**

1600 cc/day or 65 cc/hr. The formula is: 4 ml/kg/hr or 100 ml/kg/day for the first 10 kg, 2ml/kg/hr or 50 ml/kg/day for the second 10 kg, and 1 ml/kg/hr or 20 ml/kg/day for all further kg.

☐ **What is the fluid deficit in a child with mild (<5%), moderate (5-10%) and severe (>10%) dehydration?**

40-50 ml/kg, 60-90 ml/kg, and 100 ml/kg/respectively.

☐ **What are high yield criteria for clinical determination of dehydration >5%?**

Capillary refill > 2 seconds, dry mucous membranes, absent tears, and general ill appearance.

☐ **How does the urine specific gravity help you determine the level of dehydration in a child?**

<1.020 is found in mild dehydration, ~1.030 in moderate, and >1.035 in severe.

☐ **What is cardiopulmonary arrest?**

Cardiopulmonary arrest refers to an apneic, pulseless state with cardiac standstill (asystole) or a dysrhythmia producing no cardiac output (ventricular fibrillation, ventricular tachycardia).

☐ **Describe progression to non-traumatic cardiopulmonary arrest in children?**

Within the pediatric age group a cardiopulmonary arrest from a non-traumatic cause is rarely a sudden event. It occurs following an often-prolonged period of time during which a child's clinical condition deteriorates from a compensated state (respiratory distress or compensated circulatory failure) to a decompensated state (respiratory failure or decompensated shock) and finally to cardiopulmonary failure (global deficit in oxygenation, ventilation and perfusion). Once in cardiopulmonary failure, cardiopulmonary arrest is imminent.

☐ **What is the outcome of cardiac arrest in infants and children?**

The outcome of cardiac arrest in children is dismal. Most (greater than 90%) of pediatric patients who present to an Emergency Department in asystole cannot be resuscitated. Only a very small percentage of those who survive are neurologically intact.

☐ **What is shock?**

Shock is the clinical state of inadequate perfusion to meet the body's metabolic needs. Delivery of oxygen and important substrates (glucose) is impaired.

☐ **What is the difference between compensated and decompensated shock?**

In compensated shock, compensatory mechanisms (increased heart rate, increased peripheral vascular resistance) act to maintain adequate perfusion. Compensated shock is associated with an adequate blood pressure. Decompensated shock occurs when compensatory mechanisms have failed, perfusion is poor, and hypotension develops.

☐ **What three factors determine oxygen delivery?**

Cardiac output, hemoglobin concentration and oxyhemoglobin saturation.

☐ **What are clinical signs of cardiopulmonary failure?**

Respiratory failure (cyanosis, pallor), altered level of consciousness (irritability, lethargy), and hypoperfusion (tachycardia, hypotonia, weak pulses, and prolonged capillary refill). Bradycardia, hypotension and hypoventilation are seen late in cardiopulmonary failure and are signs of an imminent cardiopulmonary arrest.

☐ **How does the normal respiratory rate vary in infants and children?**

A normal respiratory rate is inversely related to a child's age. It is normally fastest in the neonate and decreases throughout infancy and childhood.

PEDIATRIC LIFE SUPPORT

☐ **What is tidal volume?**

Tidal volume is the amount or volume of air that is moved with each breath. The total tidal volume increases with age (and size), but the tidal volume/kilogram remains constant throughout life.

☐ **How is tidal volume assessed?**

Tidal volume is best assessed clinically by observation of chest wall movement (chest rise) with ventilation and by listening to air movement throughout the lungs.

☐ **What is minute ventilation?**

Minute ventilation = tidal volume x respiratory rate.

☐ **How is minute ventilation compromised with respiratory distress or failure?**

Abnormalities in the respiratory rate, tidal volume or both will compromise minute ventilation. For example, a very slow respiratory rate (bradypnea) with an adequate tidal volume or a rapid respiratory rate (tachypnea) with a very small tidal volume will both result in abnormally low minute ventilation.
MV=RR x TV
MV ⇓ when RR ⇓ as TV ⇓
MV ⇓ when RR ⇓ as TV cannot ⇑ to compensate

☐ **What should the initial assessment of respiratory function include?**

Respiratory rate (tachypnea, bradypnea, apnea), respiratory mechanics (retractions, nasal flaring, grunting, head bobbing), color of the skin and mucous membranes (cyanosis, pallor). Patients who are developing respiratory failure present with symptoms of hypoxia or hypercarbia (decreased level of consciousness, poor muscle tone, and cyanosis).

☐ **Is tachypnea always caused by respiratory problems?**

Tachypnea is a compensatory mechanism seen with metabolic acidosis, abdominal disorders and some toxic ingestion as well.

☐ **What is bradypnea?**

Bradypnea is an abnormally slow respiratory rate. An abnormally slow or decreasing respiratory rate is caused by fatigue and is an ominous sign of respiratory failure and decompensation.

☐ **What are "seesaw or rocky" respirations?**

Seesaw respirations or abdominal breathing is characterized by chest retractions accompanied by abdominal distention. It is an ineffective form of ventilation seen with severe respiratory distress and impending respiratory failure.

☐ **What is grunting?**

An ominous respiratory sound heard at the end of exhalation. It is an involuntary reflex in infants and is caused by exhalation against a prematurely closed glottis. Infants and children grunt in order to maintain or increase positive end expiratory pressure.

☐ **With what conditions is grunting associated?**

Grunting is associated with conditions that result in hypoxia such as pulmonary edema and pneumonia.

☐ **Grunting is observed during what phase of respiration?**

Expiratory-exhalation against a closed glottis.

☐ **What is stridor?**

High-pitched inspiratory sound caused by extra-thoracic upper airway obstruction.

☐ **What are common causes of upper airway obstructions in infants and young children that may present as stridor?**

Congenital anomalies (macroglossia, tracheomalacia, cysts, tumors or hemangiomas of the airway, vocal cord paralysis), infections (croup, epiglottitis, pharyngeal abscess, bacterial tracheitis), upper airway inflammation (allergy, post intubation), and esophageal or airway foreign body.

☐ **What are common pediatric diagnoses that can cause prolonged expiration?**

Bronchiolitis, asthma, intrathoracic foreign body and rarely pulmonary edema.

☐ **During positive pressure ventilation how should the effectiveness of ventilation be assessed?**

Auscultation of breath sounds and observation of chest wall movement.

☐ **What should be observed regarding chest wall expansion?**

The chest wall should expand symmetrically with each breath.

☐ **Is central cyanosis an early indicator of hypoxemia?**

No. Central cyanosis is a late sign of hypoxemia. It is generally associated with marked respiratory distress and/or respiratory failure.

☐ **Is central cyanosis always seen with hypoxemia?**

No. For cyanosis to be clinically apparent, 5 grams desaturated hemoglobin per deciliter of blood must be present. Therefore, cyanosis may not be observed in an anemic child who is hypoxic.

PEDIATRIC LIFE SUPPORT

☐ **What is the formula for estimating blood pressure in children 1-10 years of age (50th percentile)?**

80mm Hg + (child's age in years/2) mm Hg.

☐ **What compensatory mechanisms occur when shock is associated with a low cardiac output state?**

Peripheral vascular resistance increases, the skin become cool and mottled and blood is diverted away from nonessential organs, such as the skin and mesentery. This redistribution of cardiac output and increased peripheral vascular resistance allows for an adequate blood pressure to be maintained.

☐ **What occurs in shock with high cardiac output?**

In a high cardiac output state such as anaphylaxis, low peripheral vascular resistance results in increased blood flow to the skin and periphery. The skin may appear pink and well perfused and peripheral pulses may be full or bounding.

☐ **Which compensatory mechanism maintains adequate cardiac output in the face of circulatory failure?**

Heart rate to a much greater extent than stroke volume influences cardiac output in infants and young children.

☐ **What happens to the heart rate in older children in response to hypoxemia?**

The initial response to hypoxemia in older children is tachycardia. As hypoxemia persists, tachycardia is followed by bradycardia.

☐ **What compensatory mechanisms come into play to maintain normal blood pressure despite falling cardiac output?**

Tachycardia, increased cardiac contractility (does not occur in young infants) and peripheral vasoconstriction.

☐ **What is the significance of hypotension in a child with cardiopulmonary failure?**

Hypotension is a sign of endstage circulatory failure signaling depreciation of cardiovascular reserves and failure of compensatory mechanisms. It is rare but, when seen signifies an immanent cardiopulmonary arrest.

☐ **T/F: Heart rate and blood pressure are of limited value in the recognition of circulatory failure in children.**

True. Heart rate and blood pressure are of limited value in determining the presence and degree of circulatory failure.

☐ **Is blood pressure reliable for the diagnosis or estimate of degree of shock?**

Blood pressure alone is unreliable for the diagnosis or estimate of degree of shock.

☐ **What is the relationship between compensatory tachycardia and blood pressure?**

Because of compensatory tachycardia and peripheral vasoconstriction, the blood pressure is almost always found to be within the normal range in pediatric patients with progressing but still compensated circulatory failure.

☐ **As the cardiac output drops in the setting of hypovolemia, how do the peripheral pulses feel?**

The volume of the pulse seems diminished or thready. As hypoperfusion progresses, peripheral pulses may become absent.

☐ **What is the significance of the loss of central pulses in a child in cardiopulmonary failure?**

Loss of central pulses is a serious sign of an impending cardiopulmonary arrest.

☐ **What is the importance of an examination of the skin of a child who is in circulatory failure?**

Because of the high surface area to volume ratio in young children and infants, the skin is a relatively large organ as compared to adults. Decreased skin perfusion is a common and early sign of circulatory failure.

☐ **What are early signs of abnormal CNS perfusion?**

Signs of an abnormal mental status include confusion, irritability, lethargy, agitation and inappropriate behavior. An abnormal mental status in an infant or baby may be most easily assessed by observing the child's response to his parents, anxiety at parental separation and response to painful stimuli.

☐ **What are clinical signs of severe and prolonged abnormal CNS perfusion?**

More profound change in mental status, loss of consciousness, abnormal muscle tones, seizures, and pupillary dilatation. A classification of a child's level of consciousness might include awake, responsive to voice, responsive to pain, and unresponsive.

☐ **What are causes of acute deterioration during positive pressure ventilation?**

Endotracheal tube obstruction (mucus, blood, secretions), displacement (into the esophagus, pharynx or right main stem bronchus) tension pneumothorax, and mechanical equipment failure.

☐ **How does respiratory distress or failure develop during seizures?**

Poor chest wall movement occurs during generalized tonic seizures leading to hypoventilation. Upper airway obstruction may be caused by tongue displacement and secretions. The effect of the above factors can be compounded by medications that cause respiratory depression. It is important to continuously monitor a child (especially heart rate and oxygen saturation) during, and following seizures.

INFECTIOUS DISEASE

☐ **How quickly do patients infected with HIV become symptomatic?**

5-10% develop symptoms within three years of seroconversion. Symptoms of acute infection usually develop 2-4 weeks after exposure and last 2-10 weeks. This is followed by a long period of asymptomatic infection. The mean incubation time from exposure to development of AIDS is about 8.23 years for adults and 1.97 years for children < 5 years old. When AIDS develops, the survival duration is about 9 months.

☐ **An HIV+ patient presents with a history of weight loss, diarrhea, fever, anorexia, and malaise. She is also dyspneic. Lab studies reveal abnormal LFTs and anemia. What is the most likely diagnosis?**

Mycobacterium avium-complex. Treat with clarithromycin, ethambutol, and rifabutin (mycobutin).

☐ **Which drugs are used to treat CNS toxoplasmosis in AIDS patients?**

Pyrimethamine, sulfadiazine plus folinic acid.

☐ **How does acute HIV infection present?**

Fever, pharyngitis, oral ulcers, disseminated lymphadenopathy (mimics mono).

☐ **The differential diagnosis of ring enhancing lesions in AIDS patients is what?**

Lymphoma, cerebral tuberculosis, fungal infection, CMV, Kaposi's sarcoma, toxoplasmosis, and hemorrhage.

☐ **What is the presentation of an AIDS patient with tuberculous meningitis?**

Insidious onset of headache, low-grade fever, and personality changes for 2-3 weeks. This is followed by meningismus, heache, and vomiting This can ultimately lead to seizures, focal neurologic, coma, and death.

☐ **On physical exam, what is the most common eye finding in AIDS patients?**

Cotton wool spots. Identical in appearance to those in diabetes or hypertension. These lesions are believed to be incidental. However, it is important to distinguish these from early CMV infection.

☐ **What is the most common opportunistic infection in AIDS patients?**

PCP.

PEDIATRIC EMERGENCY MEDICINE BOARD REVIEW

☐ **How is Candida of the esophagus diagnosed in the ED?**

Esophagitis is typically diagnosed clinically in the ED. Endoscopy and biopsy are reserved for patients who do not respond to initial treatment with oral fluconazole or ketoconazole.

☐ **What is the risk of contracting HIV infection after an occupational exposure?**

0.3%. Eighty percent of the occupational exposure-related infections are from needle sticks.

☐ **A pediatric AIDS patient with enlargement of salivary glands, digital clubbing, and generalized lymphadenopathy and an x-ray that reveals a nodular, reticular pattern most likely has what disease?**

Lymphoid interstitial pneumonia (LIP).

☐ **Which has a worse prognosis for the pediatric AIDS patient, PCP or LIP?**

PCP.

☐ **What is the current treatment of choice for LIP therapy?**

Systemic steroid therapy.

☐ **A patient is infected with Treponema pallidum; what is the treatment?**

Treatment depends on stage. Asymptomatic infants <1 mo at risk for congenital infection may be treated with one-time dose of benzathine penicillin G 50,000 units/kg IM. Symptomatic or CNS congenital syphilis is treated with aqueous penicillin G (50,000 units/kg IV for 10 days). Stages 1° and 2° syphilis are treated with benzathine penicillin G (2.4 million units IM X 1 dose) or doxycycline (100 mg bid po for 14 day). Stage 3° syphilis is treated with benzathine penicillin G 2.4 million units IM x3 doses each at 1 week intervals.

☐ **What is the cause of chancroid?**

Hemophilus ducreyi. Patients with this condition present with one or more painful necrotic lesions. Suppurating inguinal lymphadenopathy may also be present.

☐ **What is the cause of granuloma inguinale?**

Klebsiella granulomatis. Typically the onset occurs with subcutaneous nodules that develop slowly into ulcerative or granulomatous lesions. Lesions have classic "beefy red" appearance and are located on the mucous membranes of the genital, inguinal, and anal areas.

☐ **What is the causative agent in tetanus?**

Clostridium tetani. This organism is a Gram-positive rod, vegetative, and a spore former. C. tetani produces tetanospasmin, an endotoxin, that effects inhibitory GABA and glycine receptors, leading to unopposed skeletal muscle contraction and spasm.

INFECTIOUS DISEASE

- **What is the incubation period of tetanus?**

    Variable, from 2-38 days, most cases occur between 7-10 days. Incubation is typically shorter in neonatal tetanus and in wounds closer to the CNS.

- **What is the most common presentation of tetanus?**

    Initially affects facial musculature producing trismus and risus sardonicus. Eventually involves the larger muscle groups. "Generalized tetanus" involve pain and stiffness in the trunk and jaw muscles.

- **What is the differential diagnosis for a patient who on first consideration appears to be suffering from tetanus?**

    Trismus due to dental infection, dental abscess, rabies, hypocalcemic tetany, dystonia or NMS dut to antipsychotic drugs, or the extrapyramidal effects of compazine.

- **List 3 common protozoa that can cause diarrhea.**

    1. Entamoeba histolytica. Contaminated food or water, travel to developing countries. Presents with abrupt onset of fever, abdominal pain, and bloody diarrhea; vomiting is rare. Risk of developing necrotic liver abscesses. Diagnosed by stool examination. Treatment is with metronidazole or tinidazole for acute infection followed by iodoquinol or paromomycin to prevent asymptomatic cyst passage.

    2. Giardia lamblia. Occurs worldwide, transmission from fecally contaminated food or water. It is a common cause of traverlers diarrhea. Symptoms include explosive watery diarrhea, flatus, abdominal distention, fatigue, and fever. The diagnosis is confirmed via a stool examination. Treatment in children with tinidazole or metronidazole.

    3. Cryptosporidium parvum. Fecal-oral transmission from contaminated drinking water. Symptoms are profuse watery diarrhea, cramps, N/V/F, and weight loss lasting 6-14 days. Severe infection in immunocompromised children. Treatment with Nitazoxanide.

- **Explain the pathophysiology of rabies.**

    Virally infected saliva is deposited in muscle and subcutaneous tissue where it remains for a 20-90 day incubation period. The virus eventually binds to nicotinic acetylcholine receptors and ascends to the CNS where it replicates. After replication it spreads to all peripheral tissues and organ systems.

- **What is the characteristic histologic finding associated with rabies?**

    Eosinophilic intracellular lesions found within the cerebral neurons called Negri bodies are the sites of CNS viral replication.

☐ **What are the signs and symptoms of rabies?**

Incubation period of 12 to 700 days with an average of 20 to 90 days. The initial signs and systems begin with fevers, headache, malaise, anorexia, sore throat, nausea, cough, and pain or paresthesias at the bite site.

CNS stage: agitation, restlessness, AMS, painful muscular spasms, bulbar or focal motor paresis, and opisthotonos. 20% develop ascending, symmetric flaccid and areflexic paralysis. In addition, hypersensitivity to water and sensory stimuli to light, touch, and noise may occur.

Progressive stage: lucid and confused intervals with hyperpyrexia, lacrimation, salivation, and mydriasis along with brainstem dysfunction, hyperreflexia, and extensor plantar response.

Final stage: coma, convulsions, and apnea, followed by death between 4-7 days in untreated patients. The treated patient may survive for 14 days.

☐ **How should you treat a patient with a possible rabies exposure?**

Prevention is the most effective treatment. Wound care of a suspected rabies bite should include debridement and aggressive irrigation. The wound must not be sutured; it should remain open. This will decrease the rabies infection by 90%. The wound should be infiltrated with as much RIG 20 IU/kg as possible and the remainder should be administered in the deltoid muscle. The rabies vaccine should be administered in 1 mL doses IM on days 0, 3, 7, and 14 in the opposite deltoid muscle.

☐ **In the US what is the recommended time of isolation for a dog or cat to rule out rabies?**

10 days.

☐ **In the United States, which domestic animal is most commonly infected with rabies: the cat, or the dog?**

The cat. Among wild animals, rabies are most commonly found among raccoons, skunks, foxes, coyotes, and bats.

☐ **Would a child be likely to receive a rabies vaccine following a wild rat bite?**

No. Bites from rodents rarely pose a risk of rabies.

☐ **What tick borne illness causes a systemic vasculitis and presents with a petechial rash?**

Rocky Mountain spotted fever (RMSF), it is the second most common tick borne disease. The causative agent is Rickettsia rickettsii and the vectors are the female Ixodi ticks, Dermacentor andersoni (wood tick) and D. variabilis (American dog tick). Lyme disease is the most common tick borne disease.

INFECTIOUS DISEASE

- **Where does the rash of Rocky Mountain Spotted Fever usually start?**

  On the wrists and ankles, spreading to the trunk and extremities within hours.

- **Which test should be performed in the ED to diagnose RMSF?**

  Serologic antibody testing (present 1 week after infection). You may also see thrombocytopenia, mild hyponatremia, and transaminitis. Skin biopsy with immunofluorescence is highly specific but not sensitive enough for routine diagnostic use.

- **Which antibiotics are prescribed for the treatment of RMSF?**

  Doxycycline, 2 mg/kg/dose q12h. Antibiotic therapy should not be withheld pending serologic confirmation due to risk of severe CNS infection. In this case, benefit > risk of dental staining with tetracycline.

- **Which is the most deadly form of malaria?**

  Plasmodium falciparum.

- **What is the vector for malaria?**

  The female Anopheline mosquito.

- **What lab findings are expected for a patient with malaria?**

  Normochromic normocytic anemia, a normal or depressed leukocyte count, thrombocytopenia, an elevated sed rate, abnormal kidney and LFTs, hyponatremia, hypoglycemia, and a false positive VDRL.

- **What is the drug of choice for treating P. vivax, ovale, and malariae?**

  Chloroquine.

- **Which type of parasite infections do not typically result in eosinophilia?**

  Protozoa infections, such as Amoeba, Giardia, Trypanosoma, and Babesia.

- **Which is the most common intestinal parasite in the US?**

  Giardia. Cysts are obtained from contaminated water or passed by hand to mouth transmission. Symptoms include explosive foul-smelling diarrhea, abdominal distention, fever, fatigue, and weight loss. Cysts reside in the duodenum and upper jejunum.

- **How is Chagas disease transmitted?**

  The blood sucking Reduviid "kissing" bug, blood transfusion, or breast feeding. A nodule or chagoma develops at the site of the bite. Symptoms include fever, headache, conjunctivitis, anorexia, and myocarditis. CHF and ventricular aneurysms can occur. The myenteric plexus is involved and may result in megacolon. Lab findings include anemia, leukocytosis, elevated sed rate, and ECG changes, such as PR interval, heart block, T wave changes, and arrhythmias.

PEDIATRIC EMERGENCY MEDICINE BOARD REVIEW

☐ **What are two diseases that the deer tick, Ixodes dammini, transmits?**

Lyme disease and Babesia.

☐ **How do patients present with Babesia infection?**

Intermittent fever, splenomegaly, jaundice, and hemolysis. The disease may be fatal in patients without spleens. Treatment is with clindamycin and quinine.

☐ **When are patients most likely to acquire Lyme disease?**

Late spring to late summer with the highest incidence in July.

☐ **How is Lyme disease diagnosed?**

Early lyme disease is diagnosed based on history and clinical exam significant for erythema migrans (EM). Lesion often appears prior to development of adaptive immune response. Early disseminated and late disease with diffuse EM or systemic involvement have serologic testing positive for IgM and IgG to B. burgdorferi. Treatment is amoxicillin or cefuroxime <8 yrs, or doxycycline > 8 yrs. Ceftriaxone or penicillin for CNS infection.

☐ **At which stage of Lyme disease does neurological involvement occur?**

The second and third stages.

☐ **Which is the vector responsible for the transmission of Lyme disease?**

Deer tick (Ixodes dammini).

☐ **What important feature in a patient's history should be sought when a diagnosis of Lyme disease is being considered?**

History of erythema chronicum migrans (ECM), which present in nearly 60-80% of patients early in this disease.

☐ **What is the currently recommended treatment for Lyme disease?**

For early, localized disease in children over 7 years of age, doxycycline. Amoxicillin may be used for all ages. These same drugs are used for early-disseminated disease, facial palsy, and arthritis. Recurrent arthritis, carditis and CNS disease is treated with ceftriaxone or penicillin.

☐ **How many hours must Ixodes scapularis, the deer tick, feed on a human before being able to transmit Lyme disease?**

Transmission rarely occurs if the infected tick feeds for less than 48 hours.

INFECTIOUS DISEASE

☐ **Which type of paralysis does tick paralysis cause?**

An acute symmetric, ascending, flaccid paralysis evolving over hours to days. Associated paresthesias, myalgias, restlessness, irritability. Common in children, typically resolves within 24 hrs after offending tick is removed.

☐ **What is the most common sign of tularemia?**

Ulcer at the site of the tick bite with associated painful regional adenopathy, usually cervical in children. It is caused by Francisella tularensis and is transmitted by tic vectors Dermacentor variabilis and Amblyomma americanum. Treatement with aminoglycosides.

☐ **A patient presents with a sudden onset of fever, headache, myalgias, anorexia, petechial rash. She describes the headache as retro orbital and is ex¬tremely photophobic. The patient has been on a camping trip in Wyoming. What tick-borne disease might cause these symptoms?**

Colorado tick fever, caused by a Coltivirus, found in the western mountainous US. Found in deer and porcupine, transmitted via tick D. andersoni. It is a clinical diagnosis and the disease is self limited with supportive treatment.

☐ **What is the most common cause of cellulitis?**

Staphylococcus aureus causes 80% of cellulitis (infection of the deep dermis and subcutaneous fat). Streptococcus pyogenes more commonly causes erysipelas (superficial dermis and lymphatics). Erysipelas is more commonly associated with fever and chills, butterfly facial rash and ear involvement.

☐ **What is the most common cause of cutaneous abscesses?**

Staphylococcus aureus is the most common aerobe in cutaneous abscesses; MRSA causes the majority of abscesses seen in the ED. Treatment requires incision and drainage, antibiotics are generally uneccessary in uncomplicated abscesses in healthy patients.

☐ **What is the probable cause of an infection arising from an animal bite that develops in less than 24 hours? More than 48 hours?**

Less than 24 hours is typically P. multocida or streptococci. More than 48 hours is usually Staphylococcus aureus.

☐ **A patient who was involved in a work-related crush injury to his foot is brought to your office a day after the injury in shock. The patient has a fever, severe pain, and a horrible smell coming from his foot. On palpation you feel crepitus. Diagnosis?**

Gas gangrene. Gas from the infection quickly invades the fascial planes accounting for the crepitus. Treatment is with wound debridement and systemic antibiotics (penicillin plus clindamycin).

☐ **What is the probable cause of an infection arising from an animal bite that develops in less than 24 hours? More than 48 hours?**

Less than 24 hours is typically P. multocida or streptococci. More than 48 hours is usually Staphylococcus aureus.

PEDIATRIC EMERGENCY MEDICINE BOARD REVIEW

☐ **A patient who was involved in a work-related crush injury to his foot is brought to your office a day after the injury in shock. The patient has a fever, severe pain, and a horrible smell coming from his foot. On palpation you feel crepitus. Diagnosis?**

Gas gangrene. Gas from the infection quickly invades the fascial planes accounting for the crepitus. Treatment is with wound debridement and systemic antibiotics (penicillin plus clindamycin).

☐ **What is the most common cause of gas gangrene?**

Clostridium perfringens.

☐ **What is the most common site of herpes simplex I virus infection?**

Lips and oral cavity. Often with prodrome of tingling or burning followed by eruption of painful, grouped vesicles on erythematous base. Usually heal within ten days; recurrence can be triggered by stress, sun, and illness.

☐ **What are the most common causes of otitis media?**

S. pneumoniae, nontypeable H. influenza, and M. catarrhalis.

☐ **An infant is brought to your office with fever and lethargy. On physical exam, you notice purulent rhinitis and an adherent membrane. The pt also has some shallow ulcers on the upper lip. What is your diagnosis?**

Diptheria in nares. This is more common in infants.

☐ **What are the 3 stages of pertussis? How long does each last?**

Catarrhal (1-2 weeks) paroxysmal (1-6 weeks), and convalescent (1-2 weeks).

☐ **How do you calculate dosages for antibiotics in obese children?**

Calculate their ideal weight from height and use that.

☐ **What are the most common viral causes of pneumonia in the otherwise healthy child?**

RSV, parainfluenza, less commonly adenovirus and influenza. Child is well appearing with low grade fever, symptoms of viral URI plus wheezing, crackles or rhonchi.

☐ **What viral cause of pneumonia can lead to acute fulminant pneumonia?**

Adenovirus.

☐ **What is the treatment of choice for gastroenteritis caused by Shigella in an 8-year-old child?**

TMP/SMX or azithromycin.

INFECTIOUS DISEASE

☐ **The "red man syndrome" is classically associated with what antibiotic?**

Too rapid an infusion of vancomycin.

☐ **What is the drug of choice in a child with the signs and symptoms of whooping cough?**

Azithromycin.

☐ **In a child with the typical facial features of classic mumps, where else on the body should you look for suggestive signs?**

Look for sternal edema (classic) and examine the testicles.

☐ **An 11-year-old child stepped on a nail on his way home from school. The nail pierced through his sneaker and into his foot. His tetanus status is up to date. What is your main concern?**

Infection with Pseudomonas that can lead to osteomyelitis. Pseudomonal infection is most commonly association with hot, moist environments, such as sneakers.

☐ **What would be your empiric antibiotic of choice for the boy in the preceding question?**

Ceftazidime (100mg/kg/day as TID) plus an aminoglycoside (gentamycin or tobramycin).

☐ **What is the most common cause of infectious arthritis in patients with sickle cell disease? What joint is most commonly affected?**

Staph. aureus remains the most common cause, as in otherwise healthy children. However, Salmonella is more commonly seen in septic arthritis in children with hemoglobinopathies. The hip is most commonly affected.

☐ **In a child with the clinical signs and symptoms of encephalitis, what non-invasive test can be used to determine whether herpes simplex is the causative agent?**

An EEG with high aplitude slow waves or periodic lateralized epileptiform discharges (PLEDs). An MRI of the brain is more sensitive.

☐ **If the test above is positive, what test can be used to confirm your diagnosis of herpes simplex encephalitis?**

Polymerase chain reaction (PCR) analysis of CSF.

☐ **Infection with what agent is commonly seen in children with liver transplants?**

CMV infection. Patients are prophylactically treated with ganciclovir or valcyclovir.

☐ **What is the antibiotic regimen of choice after an appendectomy in perforated appendicitis?**

Ceftriaxone or levofloxacin plus metronidazole. Antibiotics are not recommended in nonperforated appendicitis

PEDIATRIC EMERGENCY MEDICINE BOARD REVIEW

☐ **A 15-year-old boy is brought to the ED with a complaint of joint pains and general weakness. On physical exam, he has a low grade fever and hepatosplenomegaly. The father states that the boy has been really depressed lately for no known reason: "why look how happy he was just 2 weeks ago when he shot his first elk!" The father shows a picture of the son holding onto his kill. What possible diagnosis is the picture a clue to?**

It should make you consider brucellosis from handling the carcass or from ticks.

☐ **What is the easiest way to distinguish residual formula in the mouth from thrush in an infant?**

Formula is easily scraped away with a tongue depressor, while the same maneuver in a child with thrush might lead to minute bleeding points.

☐ **Cerebral calcifications are most commonly associated with what 3 congenital infections?**

Toxoplasmosis, herpes simplex and cytomegalovirus.

☐ **Minimal to severe brain dysfunction can be a sequelae of what intrauterine infections?**

Toxoplasmosis, rubella, CMV, and herpes.

☐ **Which is the most common congenital infection?**

CMV, though only about 5% will show any symptoms.

☐ **How is neonatal herpes usually contracted?**

Contact with genital secretions at delivery.

☐ **What is the treatment of choice for the neonate with an HSV infection?**

Acyclovir.

☐ **What test should be ordered to insure that a pregnant woman does not have an active hepatitis B infection?**

HBsAg.

☐ **What is the characteristic triad of manifestations for late congenital syphilis?**

Hutchinson's triad consists of Hutchinson's teeth, interstitial keratitis, and eighth nerve deafness.

☐ **What is the most effective method of reducing fever in a child?**

Acetaminophen (and NSAIDs) can return the set-point to normal.

INFECTIOUS DISEASE

☐ **What organism is responsible for most cases of occult bacteremia in infants and toddlers?**

Streptococcus pneumoniae.

☐ **A child with a positive blood culture to what organism is most likely to develop meningitis?**

Meningococcus.

☐ **How can the distribution of petechiae help one to evaluate the risk of a serious bacterial infection?**

Petechiae found only above the line of the nipples is rarely found in systemic disease.

☐ **What is necessary for the diagnosis of fever of unknown origin?**

Fever of > 38°C (101°F) for > 1 week in whom no diagnosis is apparent after initial evaluation that includes a careful history, physcial exam, and lab assessment.

☐ **Why should a child with suspected idiopathic thrombocytopenic purpura (ITP) be tested for HIV?**

Thrombocytopenia may be the presenting sign for HIV infection.

☐ **How do viral meningitis and bacterial meningitis differ with regards to CSF pressure? CSF leukocytes? CSF glucose?**

The pressure in bacterial infection is increased, whereas it is normal or slightly increased in viral. The leukocytosis is greater than 1000 (up to 60K) in bacterial, and rarely over 1000 in viral meningitis. The glucose concentration is decreased in bacterial meningitis and is generally normal in viral.

☐ **What is the most common cause of bacterial meningitis in a child greater than 2 months old?**

Strep. pneumoniae.

☐ **What antibiotic agent(s) should you start immediately in a toxic, febrile infant that is less than 3 months old?**

Cefotaxime (or gentamycin) and ampicillin.

☐ **What is the most common cause of death in children with Sickle Cell Disease in developing countries?**

Infection. Due to neonatal screening and preventative antibiotics, acute chest syndrome and multi organ failure has surpassed infection as the leading cause of death in the US.

☐ **Children with sickle cell disease are started on prophylactic penicillin by two to three months of age. This should be continued until what age?**

Routine prophylaxis with penicillin has been shown to have no effect on reduction of the risk of invasive pneumococcal infections for children older than five years of age.

PEDIATRIC EMERGENCY MEDICINE BOARD REVIEW

☐ **Children with sickle cell disease most commonly are affected with what organisms?**

Streptococcus pneumoniae, Haemophilus influenzae B, and particularly severe Mycoplasma pneumoniae infections.

☐ **What is the most common cause of infectious arthritis in patients with sickle cell disease? What joint is most commonly affected?**

Staph. aureus remains the most common cause, as in otherwise healthy children. However, Salmonella is more commonly seen in septic arthritis in children with hemoglobinopathies. The hip is most commonly affected.

☐ **Why do so many patients with meningitis become hyponatremic?**

Because a majority of patients with this disease develop some degree of SIADH.

☐ **What is the sine quo non of botulism poisoning presentation?**

Bulbar palsy.

☐ **If the mother of a child with erythema infectiosum is infected, what would be her most likely presentation?**

Arthralgia and arthritis.

☐ **Which cephalosporins cover Listeria monocytogenes?**

None. That is why ampicillin is usually added to the antibiotic regimen when infection with this organism is a possibility.

☐ **Where does the rash of Rocky Mountain Spotted Fever usually start?**

On the wrists and ankles, spreading to the trunk and extremities within hours.

☐ **A worried mother calls the ED concerned that her daughter was exposed to chickenpox at the day care center. If she were exposed, how long would it take for the symptoms to appear?**

10 - 21 days.

☐ **What is the most common cause of nosocomial bacteremia?**

Coagulase negative Staphylococcus. This is usually successfully treated with methicillin.

☐ **A child is diagnosed with impetigo from group A streptococcus. What sequelae do you have to keep an eye out for?**

Acute post-streptococcal glomerulonephritis. Impetigo does not lead to rheumatic fever.

INFECTIOUS DISEASE

☐ **What is the usual etiologic agent of a hordeolum (stye)?**

Staph. aureus.

☐ **For a hordeolum from a Staph. infection, what is the therapy of choice?**

Hot soaks, and I & D, if necessary. Routine antibiotic use is not recommended.

☐ **What is considered to be the best therapy for patients with uncomplicated cat-scratch disease?**

Symptomatic relief. Azithromycin is used for systemic or those that are immunocompromised.

☐ **A 10-year-old boy presents to your office with fever, tonsillopharyngitis, and lymphadenopathy. What laboratory tests will confirm your presumed diagnosis?**

CBC and Monospot should be all you need to confirm your suspicion of EBV (infectious mononucleosis).

☐ **What would you expect to find on a peripheral blood smear of a patient with an acquired CMV infection?**

Absolute lymphocytosis and atypical lymphocytes.

☐ **What is the drug of choice for meningococcal disease?**

Aqueous penicillin G is the ideal, though patients can be started effectively on empiric cefotaxime or ceftriaxone for suspected cases and in patients with penicillin allergy.

☐ **How often are fever and a bulging fontanelle present in an infant less than 2 months old with meningitis?**

Only about 60% have a fever, and only about 25% will have a bulging fontanelle.

☐ **Of bacterial, viral, fungal and tubercular meningitis, which typically presents with the greatest concentration of WBCs?**

Bacterial meningitis.

☐ **Meningitis due to what organism most commonly presents with a subdural effusion?**

H. influenzae.

☐ **Should people who have had contact with patients with meningococcal meningitis be given prophylactic antibiotics?**

Yes. Rifampin is recommended.

PEDIATRIC EMERGENCY MEDICINE BOARD REVIEW

☐ **What is the most common side effect of rifampin?**

Orange discoloration of urine and tears.

☐ **What is, overall, the most common cause of aseptic meningitis?**

Enteroviruses.

☐ **Generally speaking, how do exudates of viral conjunctivitis differ from bacterial conjunctivitis?**

Viral is serous, and bacterial is mucopurulent or purulent.

☐ **What is the cause of epidemic keratoconjunctivitis?**

Adenovirus.

☐ **What is the initial therapy of gonococcal ophthalmia neonatorum?**

Ceftriaxone or cefotaxime and saline irrigation of the eye until resolution of the discharge. Infants present at 2-7 days of life with conjunctival erythema, chemosis, purulent discharge. All require admission and eval for disseminated disease.

☐ **Clinically, how can you distinguish orbital cellulitis from periorbital cellulitis?**

- Orbital Cellulitis: Proptosis, pain with eye movement, impaired extraocular movement, decreased visual acuity, pupillary defect. Average age is 12. Treatment inpatient, possible surgical drainage, and 3 weeks of antibiotics.
- Periorbital Cellulitis: Erythematous, tender, indurated, swollen eyelid and periorbital area. Average age is 2. Treatment outpatient with oral amoxicillin

☐ **With a child you suspect has an otitis media from the history, but are unable to visualize the lumen secondary to cerumen obstruction, how should you proceed?**

Remove the cerumen and visualize the membrane.

☐ **What is the most common cause of hearing deficits in children?**

Otitis media. Nearly all children develop transient conductive hearing loss due to middle ear effusion during OM infection.

☐ **What is the drug of choice for streptococcal pharyngitis?**

Oral penicillin for 10 days.

☐ **After initiation of therapy for streptococcal pharyngitis, when should children be allowed back into school?**

After 24 hours of antibiotic therapy.

INFECTIOUS DISEASE

☐ **What is the classic cause of herpangina?**

Coxsackie viruses.

☐ **Does trismus more commonly occur with a peritonsillar abscess or peritonsillar cellulitis?**

Peritonsillar abscess.

☐ **What antibiotics should be given to a patient with known or suspected epiglottitis?**

Patient should receive broad spectrum antibiotics (i.e. ceftriaxone plus vancomycin) to cover S. pneumo, S. aureus, H. influenza, and b-hemolytic strep. Although widespread use of the HiB vaccine has brought the number of Haemophilus induced epiglottitis cases down dramatically, it is still present and can occur even in fully immunized children.

☐ **Are steroids effective in acute laryngotracheobronchitis?**

Yes. The use of dexamethasone (0.15-0.6 mg/kg, max 10 mg) can lead to fewer return visits, decreased length of time in ED, decreased use of epinephrine.

☐ **How often is sinus tenderness found in patients with acute bacterial sinusitis?**

Facial pain is less common in young children and sinus tenderness is rare.

☐ **What sinuses are most commonly involved in sinusitis?**

Ethmoid and maxillary sinuses. Sphenoid and frontal sinuses aren't completely formed until age 12.

☐ **What are the most common causes of acute sinusitis in children?**

Pneumococcus, H. influenzae nontypeable, and Moraxella catarrhalis (same as otitis media).

☐ **Why does therapy for TB take several months, when other infections usually clear in a matter of days?**

Because the mycobacteria divide very slowly and have a long dormant phase, during which time they are not responsive to medications.

☐ **What negative outcome can be avoided by supplementing pyridoxine in patients receiving isoniazid?**

Peripheral neuropathy, ataxia, paresthesias, and convulsions.

☐ **What are the organisms most commonly thought to be associated with Guillain-Barré Disease?**

Campylobacter jejuni (30%), CMV, EBV, and HIV.

☐ **Name at least three infectious diseases that give false positive treponemal tests (FTA, MHA-TP, TPI) for syphilis.**

Yaws, pinta, leptospirosis, rat-bite fever (Spirillum minus) Infectious mononucleosis, Lyme disease, and Malaria.

☐ **Name at least three diseases that give false positive non-treponema (VDRL, RPR) tests for syphilis.**

Mono, Chickenpox, immunizations, malignancy, IV drug use, and tuberculosis.

☐ **What is the appropriate diagnostic study to perform in order to distinguish lung sequestration from other changes on the pediatric chest x-ray?**

Angiography. Lung sequestration is a nonfunctioning mass of lung tissue that lacks normal communication with the tracheobronchial tree and receives its arterial blood supply from the systemic circulation.

☐ **Children who have cellular immunodeficiencies tend to get infections due to what organisms?**

Mycobacteria, Nocardia, CMV, varicella zoster virus, Cryptococcus, Candida, and Pneumocystis carinii.

☐ **Children with hypogammaglobulinemia tend to get infections with what organisms?**

Recurrent severe upper and lower respiratory infections (strep pneumo, h. influenza) and Pseudomonas.

☐ **Children with sickle cell disease most commonly are affected with what organisms?**

Streptococcus pneumoniae, Haemophilus influenzae B, and particularly severe Mycoplasma pneumoniae infections.

☐ **What is the evaluation for Candida of the lung?**

Fresh sputum or transtracheal aspirate should reveal yeast and pseudohyphae. Mycelia and blastospores will be seen in established colonization, so a tissue exam is needed for definitive proof.

☐ **What are some of the most convenient ways to make the diagnosis of Mycoplasma pneumoniae?**

There is no reliable diagnostic test, most are treated based on clinical suspicion and early treatment. It is possible to try antigen detection, antibody titers that are often not present at the time of presentation or DNA PCR but this is difficult distinguish from asymptomatic carriers. The cold agglutinin antibody titers is not recommended in children because accuracy is not known.

☐ **What are the most common causes of non-infectious stomatitis?**

Behcet's syndrome, Stevens-Johnson syndrome, cancer chemotherapy, and Kawasaki syndrome.

☐ **What culture media are used to grow Neisseria gonorrhoeae?**

Thayer-Martin or chocolate agar.

INFECTIOUS DISEASE

☐ **What culture medium is used to isolate Corynebacterium diphtheriae and what antibiotic should be used for treatment?**

Loeffler's medium, tellurite agar, or silica gel pack. Patients should be treated with diphtheria antitoxin and either erythromycin or penicillin.

☐ **What is the differential diagnosis of a renal abscess in a neonate?**

Renal hamartoma or congenital cyst.

☐ **What is the differential diagnosis of a renal abscess in an older infant or child?**

A Wilm's tumor, a lymphoma, an angiomyolipoma, hemorrhagic infarct, hematoma, and acute focal pyelonephritis (acute lobar nephronia).

☐ **What is the differential diagnosis for a patient who on first consideration appears to be suffering from tetanus?**

Dental abscess, rabies, hypocalcemic tetany, antipsychotic drugs, or the extrapyramidal effects of compazine.

☐ **Name at least five causes of parotitis in pediatric patients.**

An incomplete list of causes of parotitis include bacteria in general, viruses, especially mumps, echovirus, coxsackie A, lymphocytic choriomeningitis virus, parainfluenza 1 and 3, cytomegalovirus, Epstein-Barr virus, and HIV. Other causes include mycobacteria, histoplasmosis, post-typhoid fever, cat scratch disease, dehydration, collagen-vascular disease, cystic fibrosis, ectodermal dysplasia, familial dysautonomia, sarcoidosis, drugs, poisoning (including lead, copper and mercury), sialolithiasis, and tumors.

☐ **What organisms are known to cause renal and perinephric abscesses?**

The most common organisms are Staphylococcus aureus, Escherichia coli, Proteus spp., Pseudomonas spp., and Klebsiella.

☐ **What are the common bacterial causes of parotitis in newborns?**

Staphylococcus aureus, Escherichia coli, and Pseudomonas aeruginosa, and group B strep.

☐ **What are the most common etiologic agents of parotitis in older children?**

Staphylococcus aureus and streptococci.

☐ **What is the differential diagnosis of a renal abscess in a neonate?**

Renal hamartoma or congenital cyst.

☐ **What is the differential diagnosis of a renal abscess in an older infant or child?**

A Wilm's tumor, a lymphoma, an angiomyolipoma, hemorrhagic infarct, hematoma, and acute focal pyelonephritis (acute lobar nephronia).

☐ **When a child presents with more than one infected joint simultaneously, what organisms should be considered as the likely culprits?**

Staphylococcus aureus, Salmonella or gonorrhea.

☐ **The inflammation associated with tuberculous meningitis classically affects what cranial nerves?**

Cranial nerves 3,6,7, and the optic chiasm.

☐ **When a child has an underlying cardiac condition that predisposes him/her to endocarditis, what events and procedures do not require endocarditis prophylaxis?**

Dental procedures without gingival bleeding; injection of local intraoral anesthetic (except intraligamentary injections); shedding primary teeth; tympanostomy tube insertion; endotracheal intubation; flexible bronchoscopy, with and without biopsy; cardiac catheterization; gastrointestinal endoscopy, with and without biopsy; cesarean section; and, if there is no infection present, urethral catheterization; dilation and curettage; uncomplicated vaginal delivery, therapeutic abortion; sterilization procedures and insertion or removal of an intrauterine device.

☐ **Name at least five infections commonly associated with erythema nodosum.**

Erythema nodosum has been associated with many infectious and some non-infectious processes. MC causes include strep pharyngitis and idiopathic. Other common causes include sarcoidosis, tuberculosis, coccidiomycosis, histoplasmosis, Hodgkin lymphoma, chlamydia infections, blastomycosis, Behcets disease, pancreatitis, pregnancy and oral contraceptives, systemic lupus erythematosus.

☐ **Pediatric patients that have received transplantations have an increase in what infections?**

Staphylococcus, pseudomonas, Klebsiella, Candida, Aspergillus, Nocardia, Pneumocystis carinii, cytomegalovirus, and varicella zoster virus.

☐ **Children with malignancies are known to have an increase in infections due to what agents?**

Pseudomonas, Klebsiella, Escherichia coli, Cryptococcus, varicella zoster virus, Pneumocystis carinii, mycobacteria.

☐ **A child has a pericardial tap and hemorrhagic fluid is withdrawn. What are the most likely possibilities?**

Trauma, tumor, tuberculosis, or histoplasmosis.

INFECTIOUS DISEASE

☐ **What characteristics categorize a pleural fluid as empyema?**

Grossly purulent material visualized with glucose <40 mg/dL, pH <7.1, or LDH >1000 IU/L.

☐ **Name three common causes of pleural effusion without empyema in the pediatric age group.**

Staphylococcus aureus, Streptococcus pneumoniae, HiB, Streptococcus pyogenes, and Mycoplasma pneumoniae.

☐ **Name two of the most common infectious causes of an empyema in the pediatric age group.**

Staphylococcus aureus, Streptococcus pneumoniae, anaerobes, and tuberculosis.

☐ **What is the most common complication of otitis media?**

Effusion and hearing loss.

☐ **What is the most common intracranial complication of otitis media?**

Meningitis. Other intracranial complications include epidural abscess, subdural abscess, brain abscess, encephalitis, lateral sinus thrombosis, communicating hydrocephalus, CSF otorrhea, and petrositis.

☐ **What are some extracranial complication of otitis media?**

Labyrinthitis, mastoiditis, facial nerve paralysis, hearing loss, balance and motor problem, tympanic membrane perforation, cholesteatoma.

☐ **What are the indications for the prophylaxis of otitis media?**

Prophylaxis should be given after the third episode of otitis media within 6 months, or after the fourth episode of otitis media within 12 months.

☐ **How does the perforation of the tympanic membrane associated with chronic supporative otitis media differ from the perforation associated with acute otitis media?**

Tympanic membrane perforation in chronic suppurative otitis media can be permanent.

☐ **What are the most common causes of acute mastoiditis?**

Group A streptococcus, Streptococcus pneumoniae, Staphylococcus aureus, and non-typeable Haemophilus influenzae.

☐ **What are the most common intracranial complications of mastoiditis?**

Epidural empyema, brain abscess, meningitis, venous sinus thrombosis, and cerebellar abscess.

☐ **What are the most common causes of chronic mastoiditis?**

The most common causes of chronic mastoiditis are Staphylococcus aureus, and Pseudomonas aeruginosa. Less commonly other gram negative rods.

☐ **What are the most common extracranial complications of mastoiditis?**

A subperiosteal abscess, a Bezold abscess (an abscess in the submastoid space), facial nerve paralysis, temporal bone osteomyelitis and hearing loss.

☐ **What immunization is most commonly associated with post-vaccine aseptic meningitis.**

MMR

☐ **What malignancies in children are most likely to present as a fever of unknown origin?**

Most commonly caused by Leukemias and lymphomas, Other causes include neuroblastoma, hepatoma, sarcoma, and atrial myxoma.

☐ **If Creutzfeldt-Jakob disease (CJD) is a disease whose average age of presentation is 60 years, why has it suddenly become a concern in children?**

In the last 10 years, a new variant known as variant CJD (vCJD) has emerged, it is associated with bovine spongiform encephalopathy, and the consumption of contaminated beef. vCJD is distinguished from classic CJD because it affects a younger population (average age 29), and progresses less rapidly. The features of progressive dementia, ataxia, and mild clonus are present. Patients with vCJD often display psychiatric symptoms and sensory disturbances.

☐ **What is the only reliable result from a culture of urine collected by the bag method?**

No growth.

☐ **What is the most common infectious disease problem in patients with lupus who are not on steroid therapy?**

Urinary tract infections and urosepsis.

☐ **What qualifies as close contact and therefore should receive prophylaxis after contact with a patient with meningococcal disease?**

People who live in the same household, attendees of the same child care or nursery school in the previous seven days, those who have been directly exposed to the index case's secretions, such as by kissing or sharing of food, and health care providers whose mucous membranes were unprotected during resuscitation or intubation of the patient.

INFECTIOUS DISEASE

☐ **Why isn't the meningococcal vaccine in routine use in the United States?**

Serogroups B and C meningococci each cause approximately 50% of the meningococcal disease in the United States. The highest risk age group for meningococcal disease is in children under two years of age. The reason meningococcal vaccine is not routinely used in this country is because serogroup B polysaccharide is not represented in the vaccine and the serogroup C that is present is very poorly immunogenic in children under two years of age.

☐ **How often is herpes simplex virus cultured from the CSF of a pediatric patient or an adult patient with HSV encephalitis? How often is it cultured from the CSF of a neonate with HSV encephalitis?**

Herpes simplex virus may be cultured from the cerebrospinal fluid of a child six months of age or older or an adult with herpes encephalitis only 5% of the time, whereas a neonate with HSV encephalitis will grow the virus from the CSF 50% of the time.

☐ **What clinical findings are most helpful in identifying a urinary tract infection in children younger than two years**

History of UTI, Temp >40°C, suprapubic tenderness, lack of circumcision, fever >24 hrs.

☐ **Any organisms that grow in the culture of a urine specimen obtained by suprapubic aspiration can be considered to represent true infection. In symptomatic children, how many colonies in a urine culture are required to make the diagnosis when the urine specimen is obtained by catheterization?**

At least 50,000 colonies per mL defines significant bacteria, if there is 10,000 to 50,000 it is recommended that you obtain a repeat culture, if the second culture has at least 10,000 colonies per ml, with pyuria then they are considered to have a UTI.

☐ **What simple measures can be taken to prevent urinary tract infections in children?**

Taking showers instead of baths, avoiding bubble baths, treating pin worms, treating constipation and in sexually active females, post-coital voiding.

☐ **What children are candidates for influenza vaccine?**

The CDC recommends that everyone 6 months of age and older get a seasonal flu vaccine.

☐ **Do bite wounds caused by humans usually become infected with one or with multiple organisms?**

The average human bite wound contains 4 organisms per wound.

☐ **What are the most common organisms found in human bite wounds?**

Staphylococcus aureus, Streptococcus species, and Eikenella corrodens. Anaerobes are also commonly seen.

PEDIATRIC EMERGENCY MEDICINE BOARD REVIEW

☐ **In a pediatric patient who is suspected of having meningitis, cranial computed tomography (CT) of the brain should be performed before lumbar puncture under what circumstances?**

When the patient is in a coma, has papilledema or focal neurologic findings. When head CT is performed, patients should have blood cultures taken and then appropriate empiric antibiotic therapy started prior to CT.

☐ **Is it necessary to treat the male sexual partner of women with bacterial vaginosis?**

No.

☐ **If a pediatric patient with an acute illness who is receiving a third generation cephalosporin becomes secondarily infected and bacteremic, what organism would be the usual offender?**

Enterococcus.

☐ **Neonates with HSV infection may present with a localized skin, eye and mouth (SEM) infection, a disseminated infection or an encephalitis. What is the mortality rate of infants with disseminated HSV infection who are treated with acyclovir?**

29% improved from 50% due to utilization of hight dose of acyclovir (60 mg/kg/day for 21 days).

☐ **What percentage of neonatal herpes simplex virus limited to a skin, eye and mouth (SEM) disease and how are these treated?**

45% of cases are limited to superficial infection, the other 55% develop CNS or Disseminated disease. SEM disease is treated with IV acyclovir 60 mg/kg/day for 14 days, followed by oral suppressive therapy with Acyclovir 300 mg daily for 6 months. CNS or Disseminated disease is initially treated with 21 days of IV therapy before transitioning to oral suppressive therapy.

☐ **What agent is the major cause of parenterally transmitted non-A, non-B hepatitis?**

Hepatitis C virus (HCV).

☐ **What type of Haemophilus influenzae most commonly causes otitis media, conjunctivitis, and sinusitis?**

Nontypeable Haemophilus.

☐ **Does nontypeable Haemophilus influenzae ever cause pneumonia, meningitis, septicemia, endocarditis, epiglottitis, septic arthritis, post-partum bacteremia, or neonatal sepsis?**

While Haemophilus influenzae B and other typeable Haemophilus influenzae infections are more likely causes of these invasive diseases, nontypeable Haemophilus influenzae may infrequently cause them as well.

☐ **What impact does breast-feeding by the mother have on immunizations of the baby?**

None. All vaccines may be given to a baby who is breast-feeding and all vaccines may be given to a mom who is breast feeding a baby.

INFECTIOUS DISEASE

☐ **T/F: A culture from a patient is reported as growing coagulase negative staphylococci. This should be interpreted to mean Staphylococcus epidermidis.**

False. Staphylococcus epidermidis accounts for 50% of skin isolates and 75% of coagulase negative staph (CoNS) in clinical specimins. However, there are many other clinically important CoNS including S. saprophyticus a cause of UTIs in young women, and S. lugdunensis a cause of endocarditis, osteomyelitis, and sepsis.

☐ **The laboratory report reads "Streptococcus viridans." What is wrong with that report?**

There is no such thing as Streptococcus viridans. The class of organisms being referred to is rightly called Viridans streptococci. Viridans streptococci are alpha hemolytic streptococci which are composed of at least eight separate species which cause diseases in man. Examples of Viridans streptococci include: S. milleri, S. constellatus, S. anginosus and S. mutans. While most laboratories will not normally speciate a Viridans streptococci for you because they may simply be contaminants, if they are seen in two or more blood cultures or grow in pure culture from a deep site, it is appropriate to ask for speciation. The species of the organism can point to the etiology of the disease. Certain Viridans streptococci are associated with endocarditis, while others are associated with such things as carcinoma and brain abscess.

☐ **A six-year-old child presents with a history of having had a PPD placed three weeks ago which was 4 mm in induration, measured by a reliable colleague. A repeat PPD was placed one week ago and the child's forearm demonstrated 14 mm of induration. This PPD was measured by the same physician. Which PPD should be accepted as reflecting the correct amount of induration.**

The second, or "boosted" response, is accepted as the reliable response. It is routine for many institutions to test their employees with two PPD's, separated by two weeks, in order to take advantage of this phenomenon.

☐ **An eighteen month old presents with an exudative pharyngitis, fever and mildly abnormal liver function tests. You suspect mononucleosis. What study would you send?**

An EBVIgG and IGM (IgG and IgM antibody to viral capsid antigen or VCA). The chances that a mono spot will be positive in the presence of mononucleosis between 0 and 2 years of age is practically zero. The chances that a mono spot is positive between the second birthday and the fourth birthday is approximately 30%. Specific Epstein-Barr virus antibodies are much more useful for making the diagnosis in the first four years of life.

☐ **A ten month old presents with the signs and symptoms of pneumonia, which is confirmed by chest x-ray. What are the chances that this pneumonia is due to Mycoplasma pneumoniae?**

About 2%. Mycoplasma pneumoniae is hardly ever seen in the first year of life. It increases steadily as a cause of pneumonia in children into early adulthood. A convenient way to remember the chances of a child's pneumonia being due to mycoplasma is to assign a 2% chance per year of age into the twenties. By the mid-twenties, as many as 35-45% of all pneumonia is secondary to Mycoplasma pneumoniae.

☐ **Pasteurella multocida infection from an animal bite is best treated with which antibiotic?**

Penicillin VK is the drug of choice.

PEDIATRIC EMERGENCY MEDICINE BOARD REVIEW

☐ **Name two common urinary pathogens that do not give a positive urine nitrate test.**

Enterococcus and Staphylococcus saprophyticus. Acinetobacter also fails to give a positive urine nitrate test.

☐ **What unusual but important CNS side effect can be seen with ibuprofen usage?**

Aseptic meningitis.

☐ **The intervention most likely to be successful in the treatment of onychomycosis of the toenail is: a) removal of the toenail or b) antifungal therapy.**

Antifungal treatments are recommended. Oral terbinafine is first line systemic therapy, treatement duration is approx. 3-4 months. Topical treatments include efinaconazole, amorolfine, tavaborole, and ciclopirox for approx. 12-18 months, Children are thought to be better candidates for topical treatment due to thinner nail plate and faster nail growth rate.

☐ **Group B streptococcus is known to have both an early and a late onset presentation in neonatal sepsis and meningitis. What other organism does the identical thing?**

Listeria monocytogenes.

☐ **What common minor surgical procedure is associated with a decreased incidence of pyelonephritis in males?**

Circumcision.

☐ **What percentage of the time do infants with documented pyelonephritis fail to demonstrate pyuria?**

Approximately 50% of the time.

☐ **What is the incidence of occult bacteremia in infants between six and twenty-four months of age with temperature higher than 40 degrees centigrade and a white blood count >15,000/cu.mm?**

Approximately fifteen percent.

☐ **How likely is it that a child with fever and petechiae has meningococcal sepsis or meningitis?**

Seven to ten percent.

☐ **What could be concluded about the spinal fluid of a three day old infant with a CSF white count of 15 and a CSF protein of 90?**

That it is normal. At birth the CSF white count can be as high as 29. It may not fall to the expected "adult" normal of 6 or 7 until a month of age. The CSF protein can be up to 175 at birth and may take as long as three months to fall to the expected upper limit of normal of 45.

INFECTIOUS DISEASE

☐ **What is the risk of Salmonella coming into a household in an ordinary package of a dozen eggs?**

The estimated flock infection prevalence ranges from 0-35% in studied flocks. Of these, the rate of egg infection is low (0-1%) and furthermore, there is not a linear connection between the rate of egg shell infection and the rate of organism penetration inside of the egg.

☐ **What are the chances that Salmonella will be introduced into a household on contaminated meats?**

Infection control and hygiene has improved. In recent testing, 4% of chicken, 2% of ground beef, and 2% of chicken shows evidence of Salmonella contamination.

☐ **Bartonella henselae is one of the causes of Parinaud's oculoglandular syndrome (a combination of conjunctivitis and ipsilateral preauricular adenitis) and is the most common cause of chronic lymphadenopathy in children. With what animal is this organism associated?**

The cat. It is the agent of cat scratch disease.

☐ **While on the farm, Coxiella burnetii is most commonly associated with sheep. In the city a cat can aerosolize enough organisms in a closed space to be infective to people for days. What illness does Coxiella burnetii cause?**

Q Fever. This usually presents as an atypical pneumonia.

☐ **One day after sitting in a hot tub, a three-year-old child and her mother develop papular and pustular skin lesions. What is the most likely explanation?**

Pseudomonas aeruginosa dermatitis has been reported to occur in healthy individuals following immersion in hot tubs. There may be a few scattered lesions or extensive involvement. Patients may also have malaise, fever, vomiting, sore throat, conjunctivitis, rhinitis, and swollen breasts.

☐ **What are the two most common organisms known to be transmitted by unpasteurized Mexican cheese?**

Salmonella and Listeria.

☐ **What is the most productive source of positive cultures for patients with brucellosis?**

While blood, abscess and tissue cultures may be useful, bone marrow is the most sensitive and most productive source of positive cultures.

☐ **Which hemoglobin provides the greatest innate resistance to falciparum malaria?**

Erythrocytes of patients that are heterozygous for sickle cell hemoglobin (sickle cell trait) are resistant to malaria.

☐ **What is the most common infectious disease complication of both measles and influenza?**

Pneumococcal pneumonia.

PEDIATRIC EMERGENCY MEDICINE BOARD REVIEW

☐ **What rickettsial disease can be mistaken for chickenpox in inner city children?**

Rickettsialpox caused by Rickettsia akari is characterized by an initial skin lesion followed by headache, fever, chills and a vesicular rash. It is transmitted by the mouse mite.

☐ **Which rickettsial infection is most common in the United States?**

Rocky Mountain Spotted Fever caused by Rickettsia.

☐ **What is the most important therapeutic action to be taken in the teenager with infected abortion?**

Evacuation of the retained products of conception.

☐ **In the treatment of a patient with cellulitis, what therapeutic maneuver is arguably as important as antibiotic therapy?**

Elevation of the affected part.

☐ **What prophylactic antibiotic should be given systemically to burn victims?**

None. Prophylactic agents with antibacterial activity should be given topically. Xeroform gauze for minor burns, partial thickness burns can be treated with silver sulfadiazine.

☐ **What is the most likely focus of infection in a patient who presents with gram negative enteric bacteremia?**

Urinary tract infection.

☐ **What is the most likely focus of infection in a patient who presents with periorbital cellulitis unassociated with skin trauma?**

Sinusitis.

☐ **A previously healthy child has had an accidental overdose of oral iron; she appears to be septic. What is the most likely organism to cause her sepsis?**

Yersinia enterocolitica. The growth of Y. enterocolitica appears to be enhanced after exposure to excess iron. This, combined with intestinal damage to the mucosa by the iron, may play a role in pathogenesis.

☐ **A 3 month old infant presents with poor feeding, constipation, and hypotonia. There is a recent history of ingestion of honey. While you suspect infant botulism, you cannot rule out the possibility of sepsis. What class of antibiotics would you not give to this patient?**

Aminoglyosides. Gentamicin and related drugs can add to the blocked motor nerve terminals, and lead to respiratory failure in these patients with botulism.

INFECTIOUS DISEASE

☐ **T/F: It is fairly common for children to become infected with tuberculosis following casual exposures.**

False. Primary infection in children usually occurs following a prolonged close contact with an untreated adult who has cavitary disease.

☐ **A 7-year-old child is diagnosed with primary pulmonary tuberculosis and is treated appropriately with four drugs. An organism collected by gastric aspirate proves to be sensitive to three of the four drugs. Following six months of therapy, the chest x-ray appears unchanged from six months ago. What is wrong?**

Nothing. It can take many months for the chest x-ray abnormalities in pulmonary tuberculosis to show improvement. The child's symptoms, however, should have resolved and the sedimentation rate should be normal with effective therapy.

☐ **In normal children, what is the most common clinical manifestation of disease due to non-tuberculous mycobacteria?**

The most common form of non-tuberculous mycobacteria is lymphadenitis of the submandibular or anterior cervical nodes. The usual agent is M. avium intracellulare complex and M. scrofulaceum.

☐ **A teenager presents with an irritating cough which is becoming gradually worse. The boy's father had a similar problem, which was treated with erythromycin and resolved after seven days. PPD's on both the father and the son have been negative. What are the three most likely etiologic agents for this problem?**

Mycoplasma pneumoniae, Chlamydia pneumoniae, and Bordetella pertussis.

☐ **What part of the body should you examine to confirm that a vesicular rash is related to chickenpox?**

The scalp. Vesicular and crusting lesions in the scalp are typical of chicken pox. It is extremely unlikely that other causes of vesicular lesions on the skin would be associated with scalp lesions.

☐ **It is well known that children with chickenpox may eventually have an episode (or episodes) of herpes zoster. Will herpes zoster appear following an immunization with chickenpox vaccine?**

Yes. It occurs at approximately one-third to one-fifth the frequency that it does following the natural illness, but it is a distant complication of the vaccine as well.

☐ **What is the earliest age at which a child may be immunized for influenza?**

At 6 months of age.

☐ **While older infants and children with respiratory syncytial virus infection (RSV) usually present with cough and wheezing, some infants may present with what other symptom?**

Apnea.

PEDIATRIC EMERGENCY MEDICINE BOARD REVIEW

☐ **What is the name of the viral illness that presents predominantly with pharyngitis and conjunctivitis?**

Pharyngoconjunctival fever. It is caused by adenovirus.

☐ **Since only a minority of children who present with fever and petechiae have meningococcemia, what etiologic agent accounts for the bulk of the remaining cases?**

Viral causes such as enterovirus or influenza.

☐ **A ten month old infant presents with fever, diarrhea, vomiting, and mild irritability. A spinal tap shows 68 white cells, no red cells, normal protein, normal glucose, and a CSF latex agglutination that is negative for all bacterial pathogens tested. What is the most likely diagnosis?**

Rotavirus, it is estimated that up to 4% of patients with rotavirus have CNS symptoms including aseptic meningitis and seizures.

☐ **What parasitic agent should be included in the differential diagnosis when mononucleosis and CMV are considered?**

Toxoplasmosis. Everything EBV and CMV can do, toxoplasmosis can do.

☐ **When itchy lesions are seen on the palms and/or soles, what blood sucking ecoparasite should be high in the differential diagnosis?**

Sarcoptes scabiei, the etiologic agent of scabies.

☐ **Which organism is most commonly responsible for meningitis in children between 3 – 10 years of age?**

S. pneumoniae (47%), followed by N. meningitidis (32%). Treat empirically with Vancomycin and third generation cephalosporin such as Cetriaxone.

☐ **What strain of N. meningitidis is the most common cause of meningococcal disease in adolescents.**

Serogroup B accounts for 1/3 of US cases in adolescents. In 2015, the FDA approved two Serogroup B vaccines which may be administered to young adults ages 16-23, preferably between ages 16-18.

☐ **What are the common pathogens for meningitis in the neonates?**

Streptococcus group B and E. coli.

☐ **What is the most common neurologic sequel of pediatric bacterial meningitis?**

Hearing impairment.

INFECTIOUS DISEASE

☐ **What other deficits may develop following bacterial meningitis?**

Mental retardation, seizure, and spastic weakness.

☐ **What are the sequelae of viral meningitis?**

None, usually.

☐ **What are the CSF characteristics of tuberculous meningitis?**

Mononuclear pleocytosis, very high protein (several hundred to a thousand mg/dl), and very low glucose.

☐ **What is the management of herpes simplex encephalitis?**

Early treatment with acyclovir, anticonvulsants for seizures and general supportive care.

☐ **What is the most common cause of encephalitis in neonates?**

Enterovirus.

☐ **What is the recommended treatment for neurosyphilis?**

Intravenous penicillin G. Follow up CSF examinations are mandatory.

☐ **What complication may arise from aggressive penicillin treatment of neurosyphilis?**

Jarisch Herxheimer reaction.

☐ **What is Weil's disease?**

A less common variety of leptospirosis. It presents with icterus, marked hepatic and renal involvement, along with a bleeding diathesis.

☐ **How is brucellosis spread?**

Through the ingestion of contaminated milk and milk products. It may also be spread by contact with an infected animal (usually cattle), including respiratory.

☐ **What are the typical features of zidovudine associated myopathy?**

AZT is known to cause myopathy when used in high doses over prolonged periods. There is typically severe myalgia, and striking atrophy of the gluteal muscles.

☐ **Name some common causes of peripheral neuropathy in AIDS.**

Peripheral neuropathy has been postulated to be directly related to HIV, concomitant infections such as cytomegalovirus (CMV), side effects of drugs such as ddI, ddC, and 3TC, and nutritional deficiencies such as B12 deficiency.

❑ **What is the most common cause of intracranial mass lesions in HIV disease?**

Intracranial toxoplasmosis. About 20-30% of AIDS patients with positive serology for toxoplasmosis will develop toxoplasma encephalitis. The second most common intracranial mass lesion is lymphoma.

❑ **What is the current recommended treatment for intracranial toxoplasmosis in HIV disease?**

This is usually a combination therapy with sulfadiazine, pyrimethamine, and folinic acid.

❑ **What is the rationale for using folinic acid in the treatment of intracranial toxoplasmosis with pyrimethamine, and sulfadiazine combination?**

Folinic acid is thought to decrease the incidence of pyrimethamine induced hematologic toxicity

❑ **What percentage of patients with HIV disease develop CNS lymphoma?**

Approximately 2% - 6% of AIDS patients will develop primary CNS lymphoma. Up to 0.6% will present with primary CNS lymphoma concurrent with a diagnosis of AIDS.

❑ **What are the modes of transmission of HIV-1?**

Vertical: Mother to child.
Horizontal: Through sexual contact, and blood transfusion.

❑ **How is botulism contracted?**

Through the consumption of contaminated foods, by injury from non sterile objects (wound botulism), and (in infants) from intestinal colonization by Clostridium botulinum.

❑ **What are botulism's principle clinical features?**

Bilateral cranial neuropathies with symmetric descending weakness. Often with absence of fever and no sensory deficits outside of vision.

❑ **What is the organism responsible for causing Bornholm's disease?**

Bornholm disease is another name for pleurodynia during a viral illness. Coxsackieviruses, specifically group B, are responsible for this viral infection with fever and characteristic pain in the muscles of the ribs.

❑ **Name three paralytic diseases caused by an infectious agent other than polio?**

Paralytic rabies, botulism and tick paralysis.

❑ **To which group of viruses does polio belong?**

The polio virus is an enterovirus like Echovirus and Coxsackievirus.

INFECTIOUS DISEASE

☐ **What is the prophylactic regime of choice for PCP in patients with AIDS?**

Trimethoprim-sulfamethoxazole DS should be started when the CD-4+ count reaches 200 cells/mm3.

☐ **What percentage of untreated group A ß-hemolytic streptococcal infections will progress to rheumatic fever?**

Up to 3% during historic epidemics, however it is much less (<1%) when disease is endemic. Increased incidence of the disease is noted in lower socioeconomic areas.

☐ **In what age group is post streptococcal glomerulonephritis most common?**

Peak incidence is ages 5 – 12 years of age. It is uncommon in children < 3 years of age.

☐ **What is used to control outbreaks of meningococcal meningitis?**

Rifampin and ciprofloxacin are used as chemoprophylaxis for contacts.

☐ **Which has a longer incubation period in food poisoning, Staphylococcus or Salmonella?**

Salmonella. It is generally ingested in small doses, then it multiplies in the GI tract. Symptoms occur 6–48 hours after ingestion. With Staphylococcus, the disease process is dependent not on the reproduction of the organism within the host, but rather on the enterotoxin it has already produced on the contaminated food.

☐ **How many millimeters indicate a positive reaction to the Mantoux skin test in a person with HIV?**

> 5 mm. In individuals with risk factors for TB, induration must be > 10 mm. For those with no risk factors, induration must be > 15 mm to be positive.

☐ **What is the principal vector for Lyme disease in the northeastern United States?**

The black-legged tick, also known as the "deer tick", Ixodes scapularis.

☐ **An 8 year-old boy has recently acquired a pet iguana. This may place the household at increased risk of what disease?**

Salmonellosis.

☐ **What is the most common vectorborn disease in the United States?**

Lyme disease.

PEDIATRIC EMERGENCY MEDICINE BOARD REVIEW

☐ **An 8 year old male is brought in by his parents because they found a tick on his body after foing for a hike earlier that day. Does he qualify for Lyme disease prophylaxis?**

No. Prophylaxis is indicated when all the following exits:
1. The attached tick is an ixodes species and has been attached for 36 hours.
2. Local rates of infection in ticks with B. burgdorferi is > 20%.

☐ **What is the treatment of choice for prophylaxis in Lyme disease exposure?**

Doxycycline (4mg/kg) as a single dose. May use in children >8 years old. For children < 8 years of age consider amoxicillin or cefuroxime p.o.

CARDIOVASCULAR

☐ **What are the most common causes of bradycardia in a newborn?**

Hypoxemia, hypoglycemia, hypothermia, exposure to narcotics, and sepsis.

☐ **A seven day old baby is cyanotic. Physical exam reveals a holosystolic murmur. He remains cyanotic after treatment with 100% oxygen. What is the possible diagnosis and subsequent treatment of choice?**

The patient has a ductus dependent cyanotic congenital heart defect. The treatment of choice is prostaglandin E1.

☐ **When is cardioversion indicated?**

During unstable SVT and ventricular Tachycardia (VT) and ventricular fibrillation. For SVT and VT, use synchronized cardioversion.

☐ **A 5-year-old takes several KCl tablets. An ECG shows peaked T-waves, and the presence of U-waves. What is the indicated therapy?**

Calcium chloride, 10mg/kg IV; dextrose 0.5-1.0 g/kg up to 25 gms: regular insulin 0.1 U/kg IV bolus; and albuterol 5-20 mg via nebulizer.

☐ **What is the most common cause of aortic regurgitation in children?**

Aortic valve prolapse associated with a congenital ventricular septal defect.

☐ **What is the most common cause of atrial fibrillation in childhood?**

Atrial fibrillation occurs most commonly in older children with rheumatic mitral valve disease resulting in a chronically stretched atrial myocardium.

☐ **What is the mechanism that most commonly produces SVTs?**

Re-entrant tachycardias using an accessory pathway in infants, though AV node re-entry becomes more common in childhood.

☐ **What is the key feature of Mobitz I (Wenckebach) 2° AV block?**

A progressive prolongation of the PR interval until the atrial impulse is no longer conducted. If symptomatic, atropine and transcutaneous/transvenous pacing is required.

PEDIATRIC EMERGENCY MEDICINE BOARD REVIEW

☐ **Documentation of three BP results in a fully grown adolescent obtained at one week intervals must show an average greater than 140 mm Hg systolic or 80 mm Hg diastolic for the correct diagnosis of hypertension. What single reading can be considered sufficient for the diagnosis of hypertension without necessitating further investigations?**

When a patient's single reading is greater than 110 mm Hg diastolic, a diagnosis can be made. Another conclusive indicator of hypertension is when end organ damage occurs with a lower reading.

☐ **Define a hypertensive emergency in a fully grown adolescent patient.**

An elevated diastolic blood pressure that is > 115 mm Hg with associated end organ dysfunction or damage.

☐ **What are the signs and symptoms of hypertensive encephalopathy?**

Nausea, vomiting, headache, lethargy, coma, blindness, nerve palsies, hemiparesis, aphasia, retinal hemorrhage, cotton wool spots, exudates, sausage linking, and papilledema.

☐ **What is the most common cause of myocarditis in US children?**

Viruses (especially adenovirus and Coxsackie virus group B). Other causes include bacteria (diphtheria, meningococcus and tuberculosis), fungi, protozoa (Chagas' disease), spirochetes (Lyme disease), and Kawasaki disease.

☐ **What is the most common cause of mitral stenosis?**

Rheumatic heart disease.

☐ **Is mitral valve prolapse (MVP) more common in males or in females?**

Females have a stronger genetic link to the disease. 2-5% of the population has MVP.

☐ **What is the hallmark sign of mitral valve prolapse?**

A midsystolic click sometimes accompanied by a late systolic murmur. MVP is largely a clinical diagnosis. An ECG is performed to assess the degree of prolapse. Other clinical findings include a laterally displaced diffuse apical pulse, decreased S1, split S2, and a holosystolic murmur radiating to the axilla.

☐ **What physical finding is usually indicative of acute pericarditis?**

Pericardial friction rub. The rub is best heard at the left sternal border or apex with the patient in a forward sitting position. Other findings include fever and tachycardia.

☐ **Acute pyogenic pericarditis is most commonly caused by which organisms?**

Staphylococcus aureus, Neisseria meningitidis and Haemophilus influenzae type B.

CARDIOVASCULAR

☐ **What does an x-ray of a pericardial effusion reveal?**

A water bottle silhouette.

☐ **Rheumatic heart disease is the most common cause of stenosis of what 3 heart valves?**

Mitral, aortic (along with congenital bicuspid valve), and tricuspid.

☐ **Which is most likely to be pathologic in children: an S3 or an S4?**

An S4.

☐ **A short PR interval and a delta wave are characteristic of what cardiac anomaly?**

Wolff-Parkinson-White (WPW) syndrome. This is due to a reentry phenomenon along the bundle of Kent.

☐ **Which congenital heart lesions commonly present with cyanosis in an infant's first day of life?**

All the abnormalities that start with a T. Tetralogy of Fallot, Transposition of the great vessels, Total anomalous pulmonary circulation, Truncus arteriosus, and Tricuspid valve atresia (and Ebstein's malformation of the Tricuspid valve).

☐ **What is the usual presentation of children with a ventricular septal defect?**

A systolic murmur in the first days of life with no signs related to their condition. They usually resolve spontaneously.

☐ **How do infants with atrial septal defects usually present?**

The majority of these children are totally asymptomatic, and the defect usually is not diagnosed until they are school age.

☐ **A short PR interval and a delta wave are characteristic of what cardiac anomaly?**

Wolff-Parkinson-White (WPW) syndrome. This is due to a reentry phenomenon along the Bundle of Kent.

☐ **Which patients with mitral valve prolapse (MVP) should receive prophylactic antibiotics to protect against subacute bacterial endocarditis?**

Patients with MVP and a systolic murmur.

☐ **T/F: You can rule out subacute bacterial endocarditis (SBE) after you have obtained two negative blood cultures?**

False. False negatives occur 10-30% of the time in two cultures. For this reason, three blood cultures are recommended.

☐ **How does the size of a VSD correlate with the risk of infective endocarditis?**

It doesn't. The risk is independent of size.

☐ **How common do children with SBE present with cutaneous manifestations?**

Not very common. These manifestations are seen late in the disease, so they are not commonly seen in the appropriately treated patient.

☐ **What test can be used to identify vasovagal neuroregulatory syncope as the cause of syncope in children with a history of "falling out"?**

The tilt-table test.

☐ **What test can be used to differentiate congenital heart disease from primary pulmonary parenchymal disease in a newborn with cyanosis?**

The hyperoxia test. There is little to no difference in arterial oxygen content between breathing room air and 100% oxygen with cyanotic congenital heart disease

☐ **On a chest x-ray of a child, what bony abnormality would lead you to suspect coarctation of the aorta?**

Bilateral rib notching.

☐ **What bony abnormalities are seen in the chest x-rays of children with Down syndrome?**

You may see 11 ribs.

☐ **What are the common cardiac complications of juvenile rheumatoid arthritis?**

Pericarditis and (rare) myocarditis.

☐ **What are the common cardiac complications of Marfan syndrome?**

Aortic and mitral insufficiency, mitral valve prolapse (MVP), and dissecting aortic aneurysm.

☐ **What are the common cardiac complications of sickle cell anemia?**

High output cardiac failure, cardiomyopathy and cor pulmonale.

☐ **What congenital cardiac diseases will present with increased pulmonary markings on a chest x-ray?**

Transposition of the great vessels, total anomalous pulmonary venous return (and truncus arteriosus).

CARDIOVASCULAR

☐ **An "egg shaped" heart is a characteristic x-ray finding for what congenital heart malformation?**

Transposition of the great arteries.

☐ **What are the differential diagnoses for a widely split S2?**

ASD, tetralogy of Fallot (TOF), pulmonary stenosis, right bundle branch block (RBBB), total anomalous pulmonary venous return.

☐ **In classifying a heart murmur's intensity, what differentiates a III from a IV?**

The absence (III) or presence (IV) of a thrill.

☐ **What are the auscultatory findings in an ASD?**

A loud first heart sound, sometimes a pulmonic ejection click, a second heart sound that is widely split and fixed in all phases of respiration, and a systolic ejection murmur (best heard at the left middle and upper sternal border). A short, rumbling mid-diastolic murmur may be heard at the left lower sternal border.

☐ **Describe the most common 'innocent' murmur?**

Medium-pitched, "musical" or vibratory systolic ejection murmur best heard at LLSB without radiation.

☐ **How common is an 'innocent' murmur?**

Very. Up to 30% of children have one, with peak incidence between 3-7 years of age.

☐ **What is the 13th lead in a 13 lead ECG?**

V3R or V4R. These are right precordial leads which evaluate the extent of right ventricular hypertrophy.

☐ **What is the test of choice to examine for the presence of vegetations in a child with endocarditis?**

Transesophageal echocardiography.

☐ **An ECG shows the following findings: RSR' in V3R and V1, prolonged P-R interval, tall P waves, a QRS axis of -60 degrees and signs of RVH. What is your diagnosis, doc?**

AV canal defect (a.k.a. AV septal defect, endocardial cushion defect).

☐ **What is the most common cardiac malformation?**

VSD. This accounts for about 25% of all heart diseases. Prognosis depends largely on size of the defect- if small it may close spontaneously (usually before 4 years of age).

PEDIATRIC EMERGENCY MEDICINE BOARD REVIEW

☐ **A 10-year-old child with a history of VSD is about to have an impacted tooth removed. What antibiotic should he be given prophylactically to prevent endocarditis?**

Amoxicillin 50 mg/kg one hour before the procedure. If the patient has a penicillin allergy, use clindamycin 20 mg/kg one hour before the procedure.

☐ **What compound causes the ductus arteriosus to constrict right after birth?**

Oxygen, which is a vasoconstrictor.

☐ **When is it most effective to administer indomethacin to newborns with patent ductus arteriosus?**

During the first 24-48 hours of life, with a second and third dose at 12 and 36 hours, following the initial dose.

☐ **What is the treatment of choice for a child with severe aortic stenosis?**

Balloon valvuloplasty if peak systolic gradient between the LV and aorta is > 60 mmHg.

☐ **How does squatting help children with hypoxia secondary to their tetralogy of Fallot (TOF)?**

It increases their systemic vascular resistance.

☐ **You are in the ED when a newborn with severe coarctation of the aorta suddenly starts deteriorating and is acidotic. What do you do initially?**

Infuse prostaglandin E1 and administer oxygen.

☐ **What test confirms the clinical suspicion of PDA?**

2-D echocardiogram detects up to 90%.

☐ **What is the most reliable way to diagnose neonatal pulmonary hypertension?**

Right heart catheterization.

☐ **What is the normal QRS axis in a newborn?**

+135 to +180 degrees. This reflects their right ventricular dominance.

☐ **A 3-year-old child presents with premature atrial contractions (PAC's). Should this be a cause of concern?**

They are usually benign, except in children less than 1-year-old, or those on digitalis.

☐ **Name the anomalies found in tetology of Fallot (TOF).**

Pulmonary stenosis, VSD, dextroposition of aorta with septal override and RVH.

CARDIOVASCULAR

☐ **What are the most common complications of TOF prior to correction?**

Cerebral thromboses, brain abscess, bacterial endocarditis and CHF.

☐ **What is the shunting procedure currently employed to correct the cyanosis in TOF?**

The modified Blalock-Taussig shunt. This aortopulmonary shunt is a Gore-Tex conduit which anastomoses the subclavian and pulmonary arteries (homolateral branch).

☐ **What therapy can be instituted to decrease the incidence of coronary abnormalities in children with Kawasaki syndrome?**

IV immune globulin (2 g/kg given over 8 – 12 hours), and high dose aspirin.

☐ **What medication has been shown to decrease the fever and discomfort associated with an acute attack of Kawasaki syndrome?**

High dose aspirin therapy (50 – 100 mg/kg/day).

☐ **What laboratory findings differentiate Kawasaki disease from measles?**

Children with Kawasaki syndrome tend to have an increased WBC count, elevated ESR, elevated liver function and pyuria.

☐ **What congenital cardiac defects are commonly seen in children with Down syndrome?**

Endocardial cushion defect, ASD, VSD.

☐ **What congenital cardiac defects are commonly seen in infants with fetal alcohol syndrome?**

ASD, VSD.

☐ **What congenital cardiac defects are commonly seen in patient with autosomal dominant polycystic kidney disease?**

Mitral valve prolapse (MVP).

☐ **What is the 5 year survival rate of infants and children who undergo a complete heart transplantation?**

70-85% with immunosuppressive agents.

☐ **What is the primary cause of congestive heart failure (CHF) in infancy and childhood?**

Congenital heart disease.

- ☐ **A patient with congenital heart disease taking digoxin and Lasix, presents in CHF; what are the potential causes?**

    1. Untreated or unrecognized infection
    2. Anemia
    3. Arrhythmias
    4. Inadequate or excessive digitalis dose
    5. Hypokalemia
    6. Traumatic injury

- ☐ **What diseases lead to a pulsus paradoxus greater than 20 mm Hg?**

    Cardiac tamponade, asthma, CHF.

- ☐ **What does a high pulsus paradoxus indicate?**

    Circulatory compromise.

- ☐ **What are the classic presenting features of pericarditis in children?**

    Positional sharp substernal chest pain, abdominal pain, tachycardia, tachypnea, dyspnea, fever, friction rub.

- ☐ **What are the potential complications of pericarditis?**

    Myocarditis, pericardial effusion, pericardial tamponade.

- ☐ **What is the mortality rate from myocarditis?**

    35%.

- ☐ **What are the ECG findings in myocarditis?**

    1. Diffuse nonspecific ST-T wave changes.
    2. AV blocks.
    3. Ventricular ectopy.
    4. Prolonged QT interval.

- ☐ **What is the cause of sudden death in myocarditis?**

    Dysrhythmias.

- ☐ **How is acute myocarditis managed?**

    Treat the CHF, restrict fluids, diuresis, oxygen, strict bed rest, and inotropic support if needed. Treat ventricular arrhythmias with lidocaine, supraventricular tachyarrhythmias with digoxin, and complete atrioventricular block with a temporary pacemaker.

CARDIOVASCULAR

☐ **What are the ECG findings in a patient with a significant pericardial effusion?**

1. Low voltage.
2. Electrical alternans.

☐ **What condition in children is infective endocarditis most often associated with?**

Congenital heart disease (40% of cases).

☐ **What are the findings in an infant with acute CHF?**

1. Irritability
2. Poor feeding
3. Lethargy
4. Failure to thrive
5. Tachycardia
6. Tachypnea
7. Hyperactive precordium
8. Gallop rhythm
9. Rales
10. Hepatomegaly

Peripheral edema and neck vein distention are noted only in older children.

☐ **What are the most common bacteria seen in a purulent pericardial effusion?**

1. Staphylococcus aureus.
2. Streptococcus pneumoniae.
3. Haemophilus influenzae.

☐ **A patient is diagnosed with infective endocarditis. She has a history of recent dental surgery. What is the most likely organism?**

Viridans streptococcus. Viridans streptococcus is a family of organisms including such microbes of Streptococcus mutans, Streptococcus salivarius, etc.

☐ **A child with congenital heart disease develops infective endocarditis postoperatively, what is the most likely organism?**

Staphylococcal species.

☐ **In a previously healthy child without congenital heart disease, what is the most common cause of infective endocarditis.**

Staphylococcus aureus.

☐ **In what percentage of cases of bacterial endocarditis will blood cultures be positive?**

90%.

☐ **If antibiotics are required before the organism has been identified, which agents should be used?**

Aminoglycoside (gentamicin 5-7.5 mg/kg/day) + penicillinase-resistant penicillin oxacillin/ nafcillin.

☐ **A child presents with chest pain and a new murmur. Which cardiac conditions should be suspected?**

Hypertrophic obstructive cardiomyopathy and aortic stenosis.

☐ **What are the five major Jones criteria for diagnosing rheumatic fever?**

Carditis, migratory polyarthritis, chorea, erythema marginatum, and subcutaneous nodules.

☐ **Which of the major criteria is the most commonly seen?**

Polyarthritis.

☐ **Describe the rash present in rheumatic fever.**

Non-pruritic, fine, lacey in appearance with central blanching and a serpiginous pattern.

☐ **What is the most common cardiac defect in children and adolescents with rheumatic heart disease?**

Mitral insufficiency.

☐ **What are the cardiac complications of Kawasaki disease?**

Myocarditis and coronary artery aneurysms.

☐ **What percent of untreated patients with Kawasaki's disease will develop coronary artery aneurysms?**

15% to 25%. The coronary artery damage can lead to thrombosis or sudden death after acute symptoms have resolved.

☐ **What are the common causes of acquired third-degree AV block?**

Myocarditis, endocarditis, rheumatic fever, cardiac muscle disease, cardiac tumors, and postoperative cardiac surgery.

☐ **What is the most common dysrhythmia seen in the pediatric age group?**

Supraventricular tachycardia (SVT).

☐ **What is the dosage used in cardioverting a child?**

0.5 to 1 Joules/kg increasing up to 2 Joules/kg.

CARDIOVASCULAR

☐ **What is the most common cause of SVT in infants?**

Congenital heart disease (30%). Other causes include fever, infection, or drug exposure (20%), and unknown etiology (50%).

☐ **What problem can occur by using digoxin to treat SVT in a patient with the WPW syndrome?**

Digoxin can shorten the refractory period in the bypass tract and enhance conduction in the accessory pathway, leading to a rapid ventricular response and ventricular fibrillation.

☐ **What are the most common causes of atrial fibrillation and flutter?**

Congenital heart disease, rheumatic fever, dilated cardiomyopathy, hyperthyroidism.

☐ **What is the treatment of choice in an unstable patient with atrial fibrillation and flutter?**

Immediate cardioversion with 0.5 J/kg.

☐ **Why should atrial flutter be treated in a patient with congenital heart disease?**

Because this patient's risk of sudden death is four times higher than normal.

☐ **What is the mortality in untreated congenital prolonged QT syndrome?**

80%.

☐ **By what age should a child's ECG resemble an adults?**

> 3 years of age.

☐ **What are the possible etiologies for ventricular tachycardia?**

Electrolyte imbalance, metabolic disturbances, cardiac tumors, drugs, cardiac catheterization or surgery, congenital heart disease, cardiomyopathies prolonged QT syndrome, and idiopathic.

☐ **What is the normal QTc value?**

0.45 in infants, 0.44 in children, 0.43 in adolescents.

☐ **What is the treatment for premature atrial contractions (PACs)?**

No treatment is required in an asymptomatic patient but frequent PACs should be evaluated to rule out myocarditis.

☐ **When should premature ventricular contractions (PVCs) be treated?**

Treat those that cause or are likely to cause hemodynamic compromise.

PEDIATRIC EMERGENCY MEDICINE BOARD REVIEW

☐ **What is the initial treatment for symptomatic second and third degree heart block?**

Atropine 0.01 to 0.03 mg/kg. A minimum dose of 0.1 mg should be used in children to avoid atropine-induced paradoxical bradycardia.

☐ **What is the treatment for ventricular fibrillation (VF)?**

Defibrillation with 1 to 2 Joules/kg/dose.

☐ **What is the dose of epinephrine in ventricular fibrillation and asystole?**

0.1 mg/kg/dose q 3-5 minutes (1:10,000).

☐ **What is a "tet spell?"**

Episodes of paroxysmal hypoxemia which are usually self-limited and last less than 15 to 30 minutes, although they may last longer. The spells are seen more often in the morning but may occur during the day and may be precipitated by activity, a sudden fright, or injury or may occur spontaneously without apparent cause. The tachypnea and cyanosis are due to an increase in right-to-left shunting and concomitant decrease in pulmonary blood flow. The exact mechanism by which this occurs is unknown.

☐ **In the pre-term infant, what is the most common cause of congestive heart failure?**

Persistent patent ductus arteriosus.

☐ **What is the most common cause of cardiac chest pain in an infant or child?**

Anomalous pulmonary origin of a coronary artery.

☐ **What is the treatment for a "tet spell?"**

Place in the knee-chest position which increases peripheral resistance in the lower extremities, which in turn, promotes increased pulmonary blood flow. Place the patient on oxygen and give morphine sulfate 0.1 mg/kg IV. If these measures fail, use propranolol to a maximum total dose of 0.1 mg/kg.

☐ **In the pre-term infant, what is the most common cause of congestive heart failure?**

Persistent patent ductus arteriosus.

☐ **What is the most common atrial septal defect seen and what does it involve anatomically?**

Most often involves the fossa ovalis, is midseptal, and is of the ostium secundum type.

☐ **What are the most notable cardiac auscultatory findings in children with atrial septal defects?**

Normal or split first heart sound, accentuation of the tricuspid valve closure sound, a midsystolic pulmonary ejection murmur and a wide, fixed split second heart sound.

CARDIOVASCULAR

☐ **What is the typical CXR findings of a patient with atrial septal defect?**

Enlargement of the right atrium and ventricle, dilated pulmonary artery and its branches, and increased pulmonary vascular markings.

☐ **What is the confirmatory test of choice for atrial septal defect?**

Echocardiography with color Doppler flow and contrast echocardiography.

☐ **T/F: In infants and children, life-threatening cardiac rhythm disturbances are more frequently the cause rather than the result of cardiovascular emergencies.**

False. In infants and children, life-threatening cardiac rhythm disturbances are more frequently the result rather than the cause of cardiovascular emergencies.

☐ **Typically, what is cardiac arrest the result of in children?**

It is the end result of hypoxemia and acidosis resulting from respiratory insufficiency or shock.

☐ **Toward what four things must attention first be directed when confronted with life-threatening cardiac arrhythmias in children?**

Toward establishment of a patent airway, effective ventilation, adequate oxygenation, and circulatory stabilization.

☐ **Define the surface electrocardiogram.**

The surface electrocardiogram is a graphic representation of the electrical sequence of myocardial depolarization and repolarization.

☐ **What are the two key factors in influencing differences in normal heart rate in children?**

Age and physical activity.

☐ **What two primary pathological conditions may influence heart rate in children?**

Fever and volume loss (fluid loss, e.g. vomiting or diarrhea, and hemorrhage).

☐ **T/F: A febrile infant with normal cardiovascular function can have a heart rate of 200 beats per minute or higher.**

True. These infants can have heart rates up to 200 + beats per minute, heart rates in infants and young children 220 often are caused a tachyarrhythmia (supraventricular tachycardia).

PEDIATRIC EMERGENCY MEDICINE BOARD REVIEW

☐ **When should you continuously monitor the ECG in children?**

When they have evidence of respiratory or cardiovascular instability, including children who have sustained a cardiopulmonary arrest.

☐ **What is the most common cause of asystole on ECG?**

A loose electrode or wire.

☐ **What are the four most common artifacts commonly observed on ECG?**

A straight line due to a loose lead.
A tall T wave counted as an R wave by the tachometer.
Absence of P waves due to lead placement.
60-cycle interference.

☐ **When should a rhythm disturbance in a child be treated as an emergency?**

Only if it compromises cardiac output or has the potential to degenerate into a lethal (collapse) rhythm.

☐ **What is the equation for determining cardiac output?**

Stroke volume times heart rate (SV x HR).

☐ **Why can cardiac output diminish as a result of very rapid rates?**

Tachycardia may compromise diastolic filling and therefore stroke volume.

☐ **What are the two primary cardiac rhythms that result in excessively fast rates?**

Supraventricular and ventricular tachycardias.

☐ **What is the most rapid and effective method of treating tachyarrhythmias that result in cardiovascular instability and signs of shock?**

Synchronized cardioversion.

☐ **What is the cause of slow rhythms associated with cardiovascular instability in infants and children?**

AV block or suppression of normal sinus node impulse generation caused by hypoxemia or acidosis.

☐ **Define sinus tachycardia in children.**

A rate of sinus node discharge higher than normal for age.

CARDIOVASCULAR

☐ **What does the initial therapy for these slow rhythms consist of?**

Establishment of a patent airway, provision of adequate ventilation and oxygenation, and administration of medications to improve perfusion (i.e., sympathomimetics).

☐ **With what dysrhythmias is the absence of palpable pulses associated?**

Asystole, ventricular fibrillation, pulseless ventricular tachycardia, or a pulseless electrical activity, such as electromechanical dissociation.

☐ **What is the typical physiological cause of sinus tachycardia?**

The need for increased cardiac output or oxygen delivery.

☐ **Is sinus tachycardia a true arrhythmia?**

No, it is often a nonspecific clinical sign rather than a true arrhythmia.

☐ **Name six common causes of sinus tachycardia.**

Anxiety, fever, pain, blood loss, sepsis, shock.

☐ **Is sinus tachycardia regular or irregular?**

Regular, but far less regular than SVT.

☐ **Describe P wave morphology in sinus tachycardia.**

Normal, upright.

☐ **What is the therapy for sinus tachycardia?**

Therapy is directed at treating the underlying cause.

☐ **Why are attempts to decrease the heart rate by pharmacological means inappropriate in sinus tachycardia?**

Because it is a symptom, not a cause.

☐ **What is the most common cause of SVT?**

A reentry mechanism that involves an accessory pathway and/or the AV conduction system.

☐ **Which patient is least likely to tolerate SVT: small infants, infants, or older children?**

Although SVT is usually well tolerated in most infants and older children, it can lead to cardiovascular collapse and clinical evidence of shock in some small infants.

☐ **What is the usual heart rate of infants in SVT?**

Greater than or equal to 220 bpm but often higher.

☐ **What is the usual P wave morphology in SVT?**

P waves may not be visible, especially when the ventricular rate is high.

☐ **What is the QRS duration in SVT?**

QRS duration is often normal (less than 0.08 seconds) in most (>90%) children.

☐ **With what rhythm is SVT with aberrant conduction usually confused?**

Ventricular tachycardia, but this form of SVT is rare in infants and children.

☐ **What ST and T wave changes may be observed in SVT?**

ST and T wave changes consistent with myocardial ischemia may be observed if tachycardia persists.

☐ **With what rhythm is narrow complex SVT in children easily confused?**

Extreme sinus tachycardia associated with sepsis or hypovolemia, particularly because either rhythm may be associated with poor systemic perfusion.

☐ **What three things may help you distinguish SVT from sinus tachycardia?**

Heart rate, ECG, variability.

☐ **How can heart rate help you distinguish SVT from sinus tachycardia?**

Sinus tachycardia is usually associated with a heart rate less than 200 bpm. In most (60%) infants with SVT, the heart rate is greater than 230 bpm.

☐ **How can the ECG help you distinguish SVT from sinus tachycardia?**

P waves may be difficult to identify in both sinus tachycardia and SVT once the ventricular rate exceeds 200 bpm. If the P waves can be identified, the P wave axis is usually abnormal in SVT but normal in sinus tachycardia.

☐ **How can variability help you distinguish SVT from sinus tachycardia?**

In sinus tachycardia, the rate may vary from beat to beat, but there is no beat-to-beat variation in SVT. Termination of SVT is abrupt, whereas the heat rate slows gradually in sinus tachycardia.

CARDIOVASCULAR

☐ **In sum, what six indications would lead you to suspect sinus tachycardia?**

1. History compatible (fever, volume loss, hemorrhage or pain);
2. P waves present/normal;
3. HR often varies with activity;
4. Variable RR with constant PR;
5. Infants: rate usually <220 bpm;
6. Children: rate usually <180 bpm.

☐ **In sum, what six factors would lead you to suspect this was supraventricular tachycardia?**

1. History incompatible (vague or nonspecific history or history of congenital heart disease);
2. P waves absent/abnormal;
3. HR not variable with activity;
4. Abrupt rate changes;
5. Infants: rate usually >220 bpm;
6. Children: rate usually >180 bpm.

☐ **What are the first three things you should do when you encounter a child with a rapid heart rate with evidence of poor perfusion?**

1. Assess and maintain the airway;
2. Administer 100% oxygen;
3. Ensure effective ventilation.

☐ **If the QRS duration is wide for the child's age, what dysrhythmia should you treat?**

Treat as presumptive ventricular tachycardia.

☐ **Name two reversible causes of ventricular tachycardia in children.**

Electrolyte imbalance, drug toxicity.

☐ **When should vagal maneuvers be attempted in children?**

In cases of stable SVT with a pulse. Do not delay adenosine therapy or cardioversion if poor perfusion is present.

☐ **When should a 12-lead ECG be done?**

As soon as possible during treatment.

☐ **What are the indications for amiodarone?**

Recurrent v-fib, recurrent hemodynamically unstable v-tach, sustained paroxysmal atrial fibrillation, paroxysmal SVT, and atrial flutter.

PEDIATRIC EMERGENCY MEDICINE BOARD REVIEW

☐ **If rapid vascular access is available, what is your first line of treatment in symptomatic wide complex tachycardia in children?**

Amiodarone 5mg/kg IV over 20-60 minutes, or Procainamide 15mg/kg IV over 30-60 minutes, or Lidocaine 1mg/kg IV bolus (no cardioversion delays).

☐ **T/F: Amiodarone and procainamide should always be given together.**

False, do not routinely administer them together for they can lead to a prolonged QT interval.

☐ **If the antiarrhythmic doesn't work, what do you do next?**

Synchronized cardioversion.

☐ **Under what circumstance should you delay synchronized cardioversion in the symptomatic child with wide complex tachycardia?**

Never delay cardioversion.

☐ **What is the dosage for synchronized cardioversion?**

0.5-1 Joules/kg. Repeat cardioversion as needed.

☐ **If conversion with lidocaine was successful, at what rate should you start a lidocaine infusion?**

20-50 mcg/kg/minute.

☐ **A two-month-old is brought into your ER by her mother, pale and with a capillary refill time of > 2 seconds. Mom says she hasn't been acting right all day. Her heart rate is too fast to count. Monitor shows a probable SVT at 260 bpm. No IV access is available. What do you do?**

Synchronized cardioversion 0.5-1 J/kg.

☐ **You have the same patient, but with a patent IV. What do you do?**

Adenosine 0.1 mg/kg followed by rapid NS bolus 2-5ml (max dose of 0.25 mg/kg).

☐ **What is unique about the way you administer adenosine?**

Proximal port.
Both medication and flush syringe in same port.
Very rapid sequential push.
Keep your thumb on both syringes while pushing.

☐ **What do you do if the adenosine doesn't work?**

You may double the dose and repeat once.

CARDIOVASCULAR

☐ **What is the maximum dose of adenosine?**

12 mg.

☐ **A seven-year-old boy is brought by his father to your clinic. His father says that he complained of a fluttering in his chest, but otherwise feels fine. He is alert and oriented, skin is warm and dry, CRT is < 2 seconds. His pulse is strong but too rapid to count. What do you do?**

Obtain a 12-lead ECG and evaluate QRS duration, rate, and QTc interval.

☐ **You obtain a 12-lead ECG and the rate is 240 bpm. What is the normal QRS duration?**

Approximately ≤ 0.08 seconds.

☐ **The QRS is 0.05 seconds. What do you need to determine in order to treat this patient?**

Whether the rhythm is ST or SVT.

☐ **In analyzing the ECG, you notice that P waves are absent, and the rate is sustained at 240 bpm. What rhythm is this?**

Probable SVT.

☐ **The patient remains stable. What should you do?**

Establish vascular access and give adenosine.

☐ **You administer 0.1 mg/kg adenosine rapid IV push. There is no change in rhythm. What now?**

Double the dose.

☐ **Five minutes after you administer lidocaine, the rate suddenly drops to 88 and regular. The patient states the fluttering in his chest is gone. What now?**

Start a lidocaine infusion at 20-50 mcg/kg per minute. Keep waiting for the pediatric cardiologist.

☐ **A 12 month-old male with history of congenital heart disease is brought to the ED by his mother for irritability. The boy is awake and you place the child on a monitor. A rate of 200 bpm is noted and the QRS complex is .15 seconds. How will the treatment for this patient be different than the previous patient?**

Do not administer adenosine. Because of the wide QRS complex treat as if it is ventricular in origin. First, identify and treat reversible causes, such as electrolyte imbalance and drug toxicity. Then, administer Amiodarone 5 mg/kg IV over 20 to 60 minutes or Procainamide 15 mg/kg IV over 30-60 minutes or Lidocaine 1mg/kg IV. May repeat Lidocaine twice as needed.

☐ **What is the treatment of choice for patients with tachyarrhythmias (SVT, VT, atrial fibrillation, atrial flutter) who show evidence of cardiovascular compromise?**

Synchronized cardioversion.

☐ **Why is synchronized cardioversion preferable to unsynchronized countershock?**

Synchronization of delivered energy with the ECG reduces the possibility of inducing VF, which can occur if an electrical impulse is delivered during the relative refractory period of the ventricle.

☐ **What is the initial energy dose for synchronized cardioversion?**

0.5 J/kg.

☐ **What is the second energy dose for synchronized cardioversion?**

1 J/kg.

☐ **What else should you consider before cardioversion?**

Vascular access and the administration of a sedative and analgesia. However, if shock is present, cardioversion should not be delayed while these therapies are instituted.

☐ **What is adenosine?**

An endogenous purine nucleoside that blocks conduction through the AV node and interrupts AV reentry pathways.

☐ **What does adenosine cause?**

It causes transient sinus bradycardia, and usually terminates SVT rapidly, safely and effectively.

☐ **What is the half-life of adenosine?**

Less than ten seconds.

☐ **What is the duration of effect of adenosine?**

Less than two minutes.

☐ **Why does adenosine have such a short half-life?**

It is rapidly sequestered by red blood cells.

☐ **What are the primary indications for adenosine?**

It is the initial drug of choice for the diagnosis and treatment of SVT in infants and children.

CARDIOVASCULAR

☐ **Why is adenosine so effective in terminating SVT in infants and children?**

Because SVT usually involves a reentrant pathway including the AV node in these patients.

☐ **When may adenosine be ineffective in terminating SVT?**

If the arrhythmia is not due to a reentry rhythm involving the AV node or sinus node, e.g., atrial flutter, atrial fibrillation, or atrial or ventricular tachycardia.

☐ **How may adenosine be of use in arrhythmias not involving reentry at the AV or sinus nodes?**

It will not terminate the arrhythmia but may produce a transient block at the AV node that may allow the health care provider to review the 12-lead ECG and determine the underlying rhythm.

☐ **Why should you immediately push normal saline following adenosine?**

To push it quickly to the heart before the red blood cells have a chance to sequester it.

☐ **Why is it important to use two hands to administer adenosine?**

One hand on the medication syringe, one on the flush. Maintain pressure on the plunger after pushing the drug to avoid having the flush pressure push the medication back into the medication syringe.

☐ **T/F: Adenosine may not be administered intraosseously.**

False. It may be administered intraosseously if other IV access is not available.

☐ **What are the common side effects of adenosine?**

Flushing, dyspnea, chest pain, bradycardia, and irritability.

☐ **How quickly do these side effects usually resolve?**

Within 1 to 2 minutes.

☐ **With what group of patients should adenosine be used with caution?**

In children with denervated, transplanted hearts. Adenosine is contraindicated in patients taking dipyridamole (Persantine) or carbamazepine (Tegretol) because these medications can further prolong the AV block caused by the administration of adenosine.

☐ **What medications can block the receptor responsible for adenosine's electrophysiological effects?**

Therapeutic concentrations of theophylline or related methylxanthines (caffeine and theobromine).

☐ **Can you still use adenosine in a child on theophylline?**

Yes, but he or she may require a larger dose than usual.

☐ **What is verapamil?**

A calcium channel blocker.

☐ **How does verapamil exert its antiarrhythmic effect?**

By slowing conduction and prolonging the effective refractory period in the AV junctional tissue.

☐ **What effect of verapamil may cause myocardial depression?**

Its negative inotropic effect.

☐ **What negative consequences have been observed with verapamil?**

Profound bradycardia, hypotension, and asystole.

☐ **In what patients should verapamil not be used?**

It should not be used to treat SVT in an emergency setting in infants less than 1 year of age, in children with congestive heart failure or myocardial depression, in children receiving beta-adrenergic-blocking drugs, or in those who may have a bypass tract.

☐ **T/F: Ventricular tachycardia (VT) is common in the pediatric age group.**

False. It is uncommon.

☐ **What is the ventricular rate range of VT?**

The ventricular rate may vary from 120 to 400 beats per minute.

☐ **T/F: VT with a slow ventricular rate can be well tolerated.**

True.

☐ **Why is rapid VT not well tolerated?**

It compromises stroke volume and cardiac output and may degenerate into ventricular fibrillation.

☐ **Which children are most likely to develop VT?**

Those with underlying structural heart disease or prolonged QT syndrome.

CARDIOVASCULAR

☐ **What are some of the non-congenital causes of VT?**

Acute hypoxemia, hypothermia, acidosis, electrolyte imbalance, drug toxicity (e.g., tricyclic antidepressants, digoxin toxicity), and poisons.

☐ **Describe the ventricular rate for VT.**

At least 120 beats per minute and regular.

☐ **Describe the QRS in VT.**

The QRS is wide (greater than 0.08 seconds).

☐ **Describe the P waves in VT.**

P waves are not identifiable. When present, they will not be related to the QRS (AV dissociation). At slower rates the atria may be depolarized in a retrograde manner, and therefore there will be a 1:1 VA association.

☐ **Describe the T waves in VT.**

T waves are usually opposite in polarity to the QRS.

☐ **What other dysrhythmia can look like VT?**

SVT with aberrant conduction. Fortunately, aberrant conduction is present in less than 10% of children with SVT.

☐ **T/F: Previously undiagnosed wide QRS tachycardia in an infant or child should be treated as VT until proven otherwise.**

True.

☐ **How should you treat VT without palpable pulses?**

Like ventricular fibrillation (VF).

☐ **If VT is associated with signs of shock (low cardiac output, poor perfusion), but pulses are palpable, what is the indicated treatment?**

Synchronized cardioversion.

☐ **What should be done if possible prior to cardioversion?**

Intubate and ventilate with 100% oxygen, secure vascular access, provide adequate sedation and analgesia.

☐ **Why is lidocaine useful in defibrillation?**

It raises the threshold for VF.

☐ **Why should lidocaine be given prior to cardioversion?**

It suppresses postcardioversion ventricular ectopy.

☐ **When should a lidocaine drip be considered following the return of spontaneous circulation after VT or VF?**

If the ventricular arrhythmias are thought to be associated with myocarditis or structural heart disease.

☐ **To ensure adequate plasma concentration, what should you do prior to starting the infusion?**

Administer a bolus of 1mg/kg.

☐ **What effects can excessive plasma concentrations of lidocaine have?**

Myocardial and circulatory depression and possible central nervous system symptoms, including drowsiness, disorientation, muscle twitching, or seizures.

☐ **What are the most common terminal rhythms in children?**

Sinus bradycardia, sinus node arrest with a slow junctional or ventricular escape rhythm, and various degrees of AV block.

☐ **If blood pressure is normal, should you treat bradycardia?**

If there is poor systemic perfusion, it should be treated regardless of blood pressure.

☐ **What is the treatment for bradycardia?**

Adequate ventilation with 100% oxygen must be ensured, chest compressions performed, and epinephrine and atropine are administered as necessary.

☐ **What are the six possible causes of pediatric bradycardia identified by PALS?**

1. Hypoxemia;
2. Hypothermia;
3. Head injury;
4. Heart block;
5. Heart transplant;
6. Toxins/poisons/drugs.

☐ **What are some of the other causes you can think of?**

Spinal shock.
Sepsis.
Toxicity/overdose
Electrolyte imbalance.
Myocarditis.
Congenital cardiac anomaly.

CARDIOVASCULAR

☐ **Is VF a common terminal event in the pediatric age group?**

No, it is documented in only about 10% of children in whom a terminal rhythm is recorded.

☐ **What is defibrillation?**

The untimed (asynchronous) depolarization of a critical mass of myocardial cells to allow spontaneous organized myocardial depolarization to resume.

☐ **What will happen if organized depolarization does not resume after defibrillation?**

VF will continue or will progress to electrical silence, at which point restoration of spontaneous cardiac activity may be impossible.

☐ **How does synchronized cardioversion differ from defibrillation?**

Synchronized cardioversion also results in depolarization of the myocardium, but cardioversion provides depolarization that is timed (synchronous) with the patient's intrinsic electrical activity.

☐ **Why is synchronized cardioversion inappropriate in the patient with VF?**

VF has no organized cardiac electrical activity with which to synchronize.

☐ **Successful defibrillation requires the passage of sufficient electric current (amperes) through the heart. On what two factors does this current flow depend?**

The energy (joules) provided and the transthoracic impedance (ohms), which is the resistance to current flow.

☐ **If transthoracic impedance is high, what must be done to achieve sufficient current for successful defibrillation or cardioversion?**

Increase the electrical current.

☐ **What eight factors determine transthoracic impedance?**

1. Energy selected;
2. Electrode size;
3. Paddle-skin coupling material;
4. Number of shocks;
5. Time intervals between shocks;
6. Phase of ventilation;
7. Size of chest;
8. Paddle electrode pressure.

☐ **Although available information does not demonstrate a relation between energy dose and weight, pediatricians are committed to that relationship, with or without evidence! So what starting dose has been arbitrarily recommended for defibrillation?**

2 joules/kg, until further notice (or data is available).

☐ **Should you pause for CPR between shocks?**

No, unless there will be a delay due to equipment malfunction or other calamity.

☐ **After CPR, defibrillation, and epinephrine, what medications should you consider?**

Amiodarone, Lidocaine, or Magnesium.

☐ **What is important to remember to do after administering each drug?**

Circulate with good CPR for 30-60 seconds before shocking.

☐ **How soon after the administration of a medication should you shock?**

Between 30-60 seconds. Elevate the limb (if used) in which you injected the medication and perform CPR to distribute the drug prior to shocking.

☐ **What is the relationship between defibrillation pad size and impedance?**

The larger the size, the lower the impedance.

☐ **What is the optimal defibrillation pad size?**

The largest size that allows good chest contact over the entire paddle surface area and good separation between the two paddles.

☐ **May the airway person continue to hold the BVM during defibrillation?**

No, all hands must be removed from all equipment in contact with the patient, including the endotracheal tube, ventilation bag and intravenous solutions.

☐ **What is an Automated External Defibrillator (AED)?**

Automated external defibrillators are external defibrillators that incorporate a rhythm analysis system and are commonly used in adults.

☐ **What two functions are served by the adhesive pads?**

They capture the surface electrocardiogram, transmitting it through the cables to the AED unit, where it is analyzed. If a defibrillation shock is indicated, the pads provide the contact to deliver the shock to the patient.

☐ **What is a fully automated AED?**

A fully automated unit requires only that the operator apply the electrodes and turn the unit on. If the victim's rhythm is determined to be either ventricular tachycardia above a present rate or ventricular fibrillation, the unit will charge its capacitors and deliver a shock.

CARDIOVASCULAR

☐ **How does a semi-automated unit differ from one fully automated?**

It requires additional operator steps, including pressing an "analyze" button to initiate rhythm analysis and pressing a "shock" button to deliver the shock. Semi automated devices use voice prompts to assist the operator.

☐ **What shock level is delivered by most AEDs?**

200 J, although some devices have a switch to enable delivery of an alternative smaller shock (e.g. 50 J).

☐ **Can AEDs be used in pediatric arrest?**

Because current units have been developed for adults, they may be considered when the child is 8 years old or older. At this age the child weights approximately 25-30kg, so that a shock of 200 J would likely deliver a defibrillating dose of approximately 8 J/kg.

☐ **In a child 8 years old or older in cardiac arrest, what do you do until the AED arrives?**

CPR.

☐ **When is noninvasive (transcutaneous) pacing indicated in children?**

In cases of profound symptomatic bradycardia refractory to BLS and ALS.

☐ **T/F: Transcutaneous pacing has been shown to be effective in improving the survival rate of children with out-of-hospital unwitnessed cardiac arrest.**

False. It has not been shown to be effective.

☐ **Under what weight should pediatric pacing electrodes be used?**

Under 15 kg.

☐ **Where should pacing electrodes be placed on the patient prior to initiating external pacing?**

The negative electrode is placed over the heart on the anterior chest and the positive electrode behind the heart on the back. If the back cannot be used, the positive electrode is placed on the right side of the anterior chest beneath the clavicle and the negative electrode on the left side of the chest over the fourth intercostal space, in the midaxillary area.

☐ **Is precise placement of electrodes necessary to effective pacing?**

No, provided that the negative electrode is placed near the apex of the heart.

☐ **What types of pacing may be provided noninvasively?**

Either ventricular fixed-rate or ventricular-inhibited pacing.

PEDIATRIC EMERGENCY MEDICINE BOARD REVIEW

☐ **T/F: If smaller electrodes are used the pacemaker output required to produce capture generally will be lower than if larger electrodes are used.**

True.

☐ **How must pacemaker sensitivity be adjusted if ventricular-inhibited pacing is performed?**

It must be adjusted so that intrinsic ventricular electrical activity is appropriately sensed by the pacemaker.

☐ **Why is it difficult to determine if ventricular capture and depolarization are taking place?**

Because of the large pacing artifact that often occurs with transcutaneous pacing.

☐ **In this circumstance, how can you determine ventricular capture and depolarization?**

By palpating a pulse or from the pressure wave of an indwelling arterial cannula.

☐ **You are at the triage desk when mom brings in her three-month old. She states her daughter has had a fever for two days and is not feeding well. Patient is conscious and alert, skin is warm and dry. P 180, R 34, T 38, BP 88/62, SAT 98%. You place the child on a cardiac monitor and this is what you see. What is this rhythm?**

Sinus tachycardia at a rate of 180.

☐ **What are the most likely causes of this rhythm?**

Fever, anxiety, dehydration.

☐ **You are called to Room Eight to evaluate a three-week old infant. The child is conscious and alert, skin is cool, mottled and dry, capillary refill is delayed, R 64 and gasping, P 80, BP 60/40, SaO2 84%. The monitor shows the following rhythm. What is it?**

Sinus Bradycardia

☐ **What is the probable etiology of this rhythm?**

Hypoxia.

CARDIOVASCULAR

☐ **What is the immediate treatment priority?**

Administer 100% oxygen by non-rebreather mask. Assist ventilations with BVM as necessary.

☐ **You are dispatched to the scene of a high school basketball game for a player down. When you arrive, you find a fifteen-year-old player on the court. CPR is being performed by bystanders. You place your paddles in "quick look" mode and this is what you see. What is this rhythm?**

Coarse ventricular fibrillation.

☐ **What immediate action must you take?**

Rapid defibrillation.

☐ **After a shock, the patient converts to the following rhythm. What is it?**

Ventricular tachycardia.

☐ **You defibrillate at 100 joules and the patient converts to the following rhythm. What is it?**

For a fifteen year-old, this would be a sinus tachycardia if accompanied by a pulse. If no pulse, pulseless electrical activity (PEA).

☐ **Dad brings in his eight year-old son, who is conscious, alert and oriented. Dad states he was playing baseball with his son in the yard, who suddenly felt weak and dizzy. He had him rest and drink some water, but the symptoms persisted, so he brought him to the emergency department. Skin is warm and dry, PERRL, P 280, R 22, BP 110/72, SaO2 99%. The monitor shows the following rhythm. What is it?**

PEDIATRIC EMERGENCY MEDICINE BOARD REVIEW

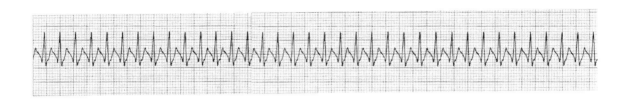

Supraventricular tachycardia.

☐ **What do you do?**

As the patient is currently stable, obtain detailed history and physical, monitor closely, establish IV access, and administer adenosine 0.1 mg/kg rapid IVP followed by a 5 ml saline flush.

☐ **How much energy should be used to cardiovert an unstable infant with a wide complex tachycardia?**

0.5–1.0 J/kg.

SEDATION AND ANALGESIA

☐ **Explain the differences between the terms Opium, Opiate, and Opioid.**

Opium: From the Greek word for juice, applies to the juice extracted from the seeds of the poppy plant, the source for over 20 alkaloids.

Opiate: Any substance that is derived from the opium plant and can induce sleep, the naturally occurring medications from the opium plant are morphine and codeine.

Opioid: Any synthetic substance not derived from opium but that can induce narcosis. Synthetic opioids are meperidine (Demerol), and fentanyl (Sublimaze).

☐ **What is the most common adverse reaction caused by local anesthetics?**

Contact dermatitis is the most common local adverse reaction and is characterized by erythema and pruritus that could progress to vesiculation and oozing.

☐ **What is the half-life of acetaminophen (Tylenol)?**

The half-life differs by age group: 7 hours in neonates, 4 hours in infants, 2-5 hours in children, 1-3 hours in adolescents, and 2-3 hours in adults. It is unaffected by renal disease as acetaminophen is metabolized in the liver.

☐ **What are the indication for naloxone (Narcan) use?**

Naloxone (Narcan) is a narcotic antagonist used to treat or prevent respiratory depression or hypotension in known or suspected narcotic overdose, neonatal opiate depression and coma of unknown origin.

☐ **How may naloxone be administered?**

Although the intravenous route is preferred, the same dosage is effective almost as rapidly following endotracheal, intramuscular or subcutaneous administration at a dose of 0.1 mg/kg /dose or 2.0 mg/dose if > 20 kg or 5 years, and can be repeated every 2-3 minutes ias needed f there is no response.

☐ **In what specific clinical situations should naloxone be used?**

Reverse CNS depression in suspected narcotic overdose, neonatal opiate depression and coma of unknown origin.

☐ **What are the contraindications to using succinylcholine (Anectine)?**

History or a family history of malignant hyperthermia, skeletal muscle myopathies, or known hypersensitivity to the drug. It should not be used in patients with a history of pseudocholinesterase deficiency.

PEDIATRIC EMERGENCY MEDICINE BOARD REVIEW

☐ **On what specific receptors in the CNS do benzodiazepines act and where are these receptors located?**

Benzodiazepines form part of the GABA receptor-chloride ion channel complex, and binding of the drug to this receptor facilitate the inhibitory action of GABA, which are excreated through increased chloride ion conductctance. The receptors are located o the thalamus, limbic structures and the cerebral cortex.

☐ **Compare the three benzodiazepines commonly used.**

The clinical effects of diazepam are shorter than lorazepam because it dissociates more rapidly from receptor sites than diazepam, however the elimination half-lives are longest for diazepam.

Medication	Elimination half time in minutes
diazepam	20-35
lorazepam	10-20
midazolam	1-4

☐ **Describe the state induced with the use of ketamine (Ketalar)?**

Ketamine produces a state of dissociative anesthesia characterized by analgesia, normal pharyngeal-laryngeal reflexes, normal skeletal muscle tone, cardiovascular and respiratory stimulation, and occasional transient, respiratory depression.

☐ **What is the most commonly sited contraindication for ketamine use?**

Ketamine is contraindicated in patients in whom a significant elevation of blood pressure would constitue a serious hazard including those with head trauma, intracranial mass, or intracranial bleeding.

☐ **How is intravenous ketamine administered for induction in a pediatric patient?**

1-2 mg/kg per initial IV, given over 60 seconds, produces 5-10 minutes of anesthesia, and used in increments of 1/2 of an initial dose if needed to maintain anesthesia.

☐ **What are the common symptoms during recovery from intravenous ketamine?**

Common symptoms in patients recovering from IV ketamine include emergence delirium often consisting of dissociative or floating sensations. These reactions occur more frequently in patients between 15 and 45 years of age and typically last only a few hours, although some patients may experience flashbacks several weeks later.

☐ **How can the side effects of ketamine be lessened during recovery?**

The concomitant use of a benzodiazepine can lessen the side effects commonly seen during emergence.

☐ **What is the dose and route of administration of fentanyl (Sublimaze) when used in the neonate?**

In the neonate, fentanyl can be given IV or IM at a dose of 1-2 micrograms/kg infused slowly over 1 minute.

SEDATION AND ANALGESIA

☐ **What is the half-life of morphine?**

The half life of morphine is approximately 9 hours in preterm neonates, 6.5 hours in term neonates aged 0 to 57 days and 2 hours in infants and children aged 11 days to 15 years.

☐ **Describe the pharmacokinetics of codeine.**

Codeine has adequate oral absorption and is metabolized in the liver to morphine. It has a half-life of 3-4 hours and is excreted in the urine.

☐ **Why should IV fentanyl be infused slowly?**

Rapid IV infusion of fentanyl can result in skeletal muscle and chest wall rigidity, leading to impaired ventilation, respiratory distress and eventual apnea.

☐ **How potent is fentanyl, when compared to morphine?**

Fentanyl is 100 times more potent.

☐ **What are the contraindications to using eutectic mixture of local anesthetics (EMLA) cream?**

Contraindications include known hypersensitivity to lidocaine (Xylocaine), prilocaine, or local anesthetics of the amide type. EMLA should be used with caution in patients with idiopathic or congenital methemoglobinemia.

☐ **What is the mechanism of action of acetaminophen?**

Acetaminophen inhibits the synthesis of prostaglandins in the central nervous system and peripherally blocks pain impulse generation. It also produces antipyresis by inhibiting the hypothalamic heat regulating center.

☐ **Why should codeine not be given intravenously?**

When given intravenously, codeine causes a large histamine release which leads to palpitations, hypotension, bradycardia, and peripheral vasodilatation.

☐ **Compare the cardiovascular adverse effects of fentanyl when compared to morphine.**

Fentanyl has less hypotensive effects than morphine due to minimal or no histamine release.

☐ **Is it appropriate to use eutectic mixture of local anesthetics on mucous membranes?**

No, because it has increased absorption across mucous membranes which can lead to toxic levels. In addition, EMLA application to oral mucosa can interfere with swallowing and increase the risk of aspiration. Patients should not ingest food for at least 1 hour after the use of anesthetic agents in the mouth or throat.

☐ **What is the maximum daily dose of ibuprofen (Motrin, Advil)?**

40 mg/kg, up to a 3 grams maximum.

PEDIATRIC EMERGENCY MEDICINE BOARD REVIEW

☐ **What are some of the misconceptions about pain in children?**

The pain perception in children is decreased because of biologic immaturity and children have a higher tolerance to pain. Other misconceptions are that children have little or no memory of pain and that not only are they more sensitive to the side effects of analgesics, but that they are at special risk for addiction to narcotics. All of these beliefs are false.

☐ **Discuss pain conduction in the neonate.**

In the neonate, as in the adult, unmyelinated C fibers transmit nociceptive information peripherally. Nerve pulse transmission in incompletely myelinated A-d fibers is delayed, not blocked, until myelination has been completed postnatally. The shorter distances necessary for impulse travel offset any delay in conduction velocity.

☐ **Name 3 common paralytic non-depolarizing agents?**

Pancuronium (Pavulon), vecuronium (Norcuron), and Rocuronium (Zemuron).

☐ **Describe the mechanism of action of non-depolarizing paralytic agents?**

The non-depolarizing neuromuscular blockers compete with acetylcholine for cholinergic receptor sites on the postjunctional membrane. When they bind they inhibit the transmitter action of acetylcholine. They do not relieve pain and do not effect consciousness.

☐ **What may be found in the urine of prepubescent children after administration of succinylcholine?**

Myoglobin may be found in the urine due to muscle fasciculations.. About 40% of prepubertal children exhibit this effect; however, it is not related to the apparent severity of the fasciculations.

☐ **How long does anesthesia last from a perineural injection of lidocaine?**

When used alone it lasts 60-75 minutes and when used in combination with epinephrine it lasts up to two hours.

☐ **How long does it take for a local injection of lidocaine to exert its effect?**

Its effect is immediate.

☐ **What are the routes of administration of chloral hydrate?**

Never intravenous. Chloral hydrate may only be given PO or PR.

☐ **What are the contraindications to using chloral hydrate?**

Chloral hydrate is contraindicated in patients with severe cardiac disease and in patients with hepatic or renal impairment.

SEDATION AND ANALGESIA

☐ **What is the maximum dose of locally injected lidocaine?**

7.0 mg/kg/dose with epinephrine every 2 hours or 4.0-5.0 mg/kg/dose without epinephrine every 2 hours.

☐ **Describe the mechanism of action of the anesthetic agent propofol (Diprivan).**

Propofol is chemically unrelated to any other anesthetic drugs and it most closely resembles the alcohol family. It is a short-acting, lipophilic sedative/hypnotic that has the potential to cause global CNS depression through agonist actions on GABAa receptors. It is anecdotally called milk of amnesia. It may cause apnea.

PULMONARY

☐ **What should you suspect if a patient has symptoms similar to those associated with emphysema but he/she is very young and is not a smoker?**

Alpha-1-antitrypsin deficiency. Without alpha-1-antitrypsin, excess elastase accumulates resulting in lung damage.

☐ **A 14-year-old high school freshman attended three days of classes at his new high school last week before it was shut down because of the potential of asbestos exposure. The boy currently exhibits a non-productive cough and says his chest hurts. Does this boy have asbestosis?**

Although a non-productive cough and pleurisy are symptoms of asbestosis, other signs, such as exertional dyspnea, malaise, clubbed fingers, crackles, cyanosis, pleural effusion and pulmonary hypertension, should be displayed before making a diagnosis of asbestosis. In addition, asbestosis does not develop until 10-15 years after regular exposure to asbestos.

☐ **What should be done to decrease aspiration in a patient who is continuously vomiting and at risk for aspiration pneumonia?**

Lie the patient on his right side in Trendelenburg. This will help confine the aspirate to the right upper lobe.

☐ **How can aspiration be prevented when intubating?**

By appropriate use of rapid sequence induction, including minimizing unnecessary ventilation and applying cricoid pressure during intubation.

☐ **Where does aspiration generally occur as revealed by chest x-ray?**

The lower lobe of the right lung. This is the most direct path for foreign bodies in the lung.

☐ **What pH of aspirated fluid suggests a poor prognosis?**

pH < 2.5.

☐ **A 16-year-old triathlete develops wheezing when exercising. Describe a relevant treatment program for this individual.**

Beta agonist prior to exercise is the initial management therapy.

☐ **What x-ray findings may be present in a patient with a long history of bronchial asthma?**

Increased bronchial wall thickening (caused by epithelial inflammation) and flattening of the diaphragm.

PEDIATRIC EMERGENCY MEDICINE BOARD REVIEW

☐ **What pulmonary function test is the most diagnostic for asthma?**

FEV1/FVC. The amount of air exhaled in 1 second, in comparison to the total amount of air in the lung that can be expressed, is determined by performing this test. A ratio of under 80% is diagnostic of asthma. Peak flow monitors are helpful in monitoring asthma at home.

☐ **Is wheezing a required finding in asthma?**

No. Thirty-three percent of children with asthma will have only cough variant asthma with no wheezing.

☐ **Which is more effective for relieving an acute exacerbation of bronchial asthma in a conscious patient, nebulized albuterol or albuterol MDI administered via an aerosol chamber?**

They are equally efficacious when performed appropriately.

☐ **A child is born to two parents who are both asthmatic. What is the probability of their child having asthma?**

Up to 50%.

☐ **What are the most common extrinsic allergens that affect asthmatic children?**

Dust and dust mites.

☐ **What treatment should be initiated for a patient with acute asthma who does not improve with humidified O2, albuterol nebulizers, steroids, or anticholinergics?**

In the setting of respiratory distress: Subcutaneous epinephrine, 0.3 cc of 1:1,000. (0.01 cc/kg to a max of 0.3 cc)

☐ **What kind of medical family history are asthmatic patients most likely to have?**

Asthma, seasonal allergies, and/or atopic dermatitis.

☐ **Which type of exercise usually triggers an asthmatic episode in patients with exercise-induced asthma?**

High intensity exercise for more than 5-6 minutes.

☐ **What is the role of salmeterol in the asthmatic patient?**

It is used for the long term management of these patients and not in the acute setting. Because of its long half-life, it may be ideal for chronic use in nocturnal break-through exacerbations.

PULMONARY

☐ **What effects do increased levels of cAMP have on bronchial smooth muscle and on the release of chemical mediators, such as histamine, proteases, platelet activation factor and chemotactic factors, from airway mast cells?**

Relaxes smooth muscle and decreases the release of mediators. Recall that the effects of cAMP are opposed by cGMP. Thus, another treatment approach can be provided by decreasing the levels of cyclic GMP via the use of anticholinergic (antimuscarinic) agents, such as ipratropium bromide.

☐ **As discussed above, stimulation of ß-adrenergic receptors increases cAMP availability and results in smooth muscle relaxation. Which flavor of ß-adrenergic receptors primarily control bronchiolar and arterial smooth muscle tone?**

ß2-adrenergic receptors.

☐ **What effect does acidosis have on treatments with alpha-adrenergic agonists?**

Decreased efficacy.

☐ **If corticosteroids are prescribed, how should prednisone or prednisolone be dosed?**

1-2 mg/kg/day, may be divided into two divided doses or once a day for 5 days. No tapering is necessary if duration of therapy is 5 day or less.

☐ **What is the appropriate parenteral dose of methylprednisolone (Solu-Medrol) to administer to a pediatric patient with status asthmaticus?**

1-2 mg/kg every 6 hours.

☐ **If mechanical ventilation is required for such a patient, what is an appropriate setting for the initial tidal volume?**

7 - 10 mL/kg (goal is to keep pressure < 50 mmHg).

☐ **Bronchiectasis occurs most frequently in patients with what conditions?**

Patients with cystic fibrosis, immunodeficiencies, lung infections, or foreign body aspirations.

☐ **Which is the most common type of pathogen in bronchiolitis?**

RSV (respiratory syncytial virus). It generally affects infants younger than 2 years old.

☐ **A 2-year-old patient presents with a sudden harsh cough that is worse at night, wheezing, and rhonchi/inspiratory stridor. What is the diagnosis?**

Croup (also known as laryngotracheitis). This condition is preceded by URI symptoms, and is most frequently caused by the parainfluenza virus. A barking seal cough is a characteristic of croup.

PEDIATRIC EMERGENCY MEDICINE BOARD REVIEW

☐ **In treating a patient with a common cold, you prescribe an oral decongestant. Is it necessary to also suggest an antitussive?**

Most coughs arising from a common cold are caused by the irritation of the tracheobronchial receptors in the posterior pharynx by postnasal drip. Postnasal drip can be cleared up with decongestant therapy, thereby eliminating the need for cough suppressant therapy.

☐ **What are the most common etiologies of a chronic cough?**

Postnasal drip (40%), asthma (25%), and gastroesophageal reflux (20%). Other etiologies include bronchitis, bronchiectasis, bronchogenic carcinoma, esophageal diverticula, sarcoidosis, viruses, and drugs.

☐ **Which condition must be ruled out when childhood nasal polyps are found?**

Cystic fibrosis.

☐ **What age group is afflicted with the most colds per year?**

Kindergartners win the top billing with an average of 12 colds per year. Second place goes to preschoolers with 6-10 per year. School children get an average of 7 per year, and adolescents and adults average only 2-4 per year.

☐ **What is the duration of a common cold?**

3-10 day self-limited course.

☐ **Which age group usually contracts croup?**

6 months to 3 years. Croup is characterized by cold symptoms, a sudden barking cough, inspiratory and expiratory stridor, and a slight fever.

☐ **A newborn presents with poor weight gain, steatorrhea, and a GI obstruction arising from thick meconium ileus. What test should be performed?**

The "sweat test," which detects electrolyte concentrations in the sweat. The infant may have cystic fibrosis. Cystic fibrosis is an autosomal recessive defect affecting the exocrine glands. As a result, electrolyte concentrations increase in the sweat glands. Genetic testing is also available.

☐ **What is the classic triad of cystic fibrosis?**

1. Chronic pulmonary disease.
2. Pancreatic enzyme deficiency.
3. Abnormally high concentration of sweat electrolytes.

☐ **What is the most common presentation of newborns with cystic fibrosis?**

GI obstruction due to meconium ileus.

PULMONARY

☐ **What is the difference between the "cough of croup" and the "cough of epiglottitis"?**

Croup has a seal like "barking" cough, while epiglottitis is accompanied by a minimal cough. Children with croup have a hoarse voice while those with epiglottitis have a muffled voice.

☐ **What is the "thumb print sign"?**

A soft tissue inflammation of the epiglottis on a lateral x-ray of the neck.

☐ **Are aspirated foreign bodies more likely to be found in the right or left bronchus?**

Right.

☐ **What is the definition for massive hemoptysis?**

Coughing, not vomiting, of more than 300 mL of blood in 24 hours.

☐ **Life threatening hemoptysis should be suspected when there is:**

1. A large volume of blood.
2. The appearance of a fungus ball in a pulmonary cavity via chest x-ray.
3. Hypoxemia.

☐ **What are Kerley B lines?**

X-ray findings found in pulmonary edema. Fluid in the lung causes waterlogged interlobular septa that become fibrotic with time. These show up as lines and shadows on an x-ray.

☐ **Which type of bacteria is most commonly found in lung abscesses?**

Anaerobic bacteria.

☐ **Which single antibiotic is the most effective for treating uncomplicated lung abscess?**

Clindamycin.

☐ **T/F: Flora of lung abscesses are usually polymicrobial.**

True.

☐ **Which age group is usually afflicted by pertussis?**

Infants younger than 1.

☐ **Where does pain from pleurisy radiate?**

The shoulder, as a result of diaphragmatic irritation.

PEDIATRIC EMERGENCY MEDICINE BOARD REVIEW

☐ **Which is the most common community-acquired pneumonia?**

Pneumococcal pneumonia.

☐ **Pneumatoceles, thin-walled air-filled cysts, on an infant's x-ray are a sign of which type of pneumonia?**

Staphylococcal pneumonia.

☐ **What are the most common causes of Staphylococcal pneumonias?**

Drug use and endocarditis. This pneumonia produces high fever, chills, and a purulent productive cough.

☐ **What are the extrapulmonary manifestations of mycoplasma?**

Erythema multiforme, pericarditis, and CNS disease.

☐ **What is the most frequent etiology of nosocomial pneumonia?**

Pseudomonas aeruginosa. It most frequently occurs in immunocompromised patients or patients on mechanical ventilation. Other common causes include viral infections, staphylococcus, enterobacter, and anaerobes.

☐ **A 14-year-old comes to your office bragging, between coughs, about a 3 day Young Republicans convention she attended last week in Las Vegas. She is nauseous and coughing; she has chills and a fever of 103.5°F. She also has a minor gait problem and splints her chest when she breaths. Should you be concerned about your own health as she has just given a mighty cough in your direction?**

This patient most likely has Legionnaires' disease (legionellosis). Unless you attended the same convention in Las Vegas, you are unlikely to catch this pneumonia. Legionella pneumophila contaminates the water in air conditioning towers and moist soil. It is not spread from person to person.

☐ **Describe the classic chest x-ray findings in a patient with mycoplasma pneumonia.**

Patchy diffuse infiltrates are the most common. Pneumatoceles, cavities, abscesses, and pleural effusions can occur, but are uncommon.

☐ **Describe the classic chest x-ray findings in Legionella pneumonia.**

Multifocal, bilateral consolidations and a bulging fissure. Radiologic changes may lag behind the clinical picture.

☐ **Describe the different presentations of bacterial and viral pneumonia.**

Bacterial pneumonia is typified by a sudden onset of symptoms, including pleurisy, fever, chills, productive cough, tachypnea, and tachycardia. The most common bacterial pneumonia is pneumococcal pneumonia. Viral pneumonia is characterized by gradual onset of symptoms, no pleurisy, chills or high fever, general malaise, and a non-productive cough.

PULMONARY

☐ **A 17-year-old college student is home for winter break and presents complaining of a 10 day history of a non-productive dry hacking cough, malaise, a mild fever, and no chills. What is the diagnosis?**

Mycoplasma pneumoniae, also known as walking pneumonia. Although this is the most common pneumonia that develops in teenagers and young adults, it is an atypical pneumonia and is most frequently found in close contact populations (e.g., schools and military barracks).

☐ **Match the pneumonia with the treatment.**

1. Klebsiella pneumoniae
2. Streptococcal pneumoniae.
3. Legionella pneumophila.
4. Haemophilus influenza pneumonia.
5. Mycoplasma pneumoniae.

a) Erythromycin, azithromycin, tetracycline, or doxycycline.
b) Penicillin G.
c) Cefuroxime and clarithromycin.
d) Erythromycin and rifampin.

Answers: (1) c, (2) b, (3) d, (4) c, and (5) a.

☐ **What is the most common cause of pneumonia in children?**

Viral pneumonia. Infecting viruses include influenza, parainfluenza, RSV, and adenoviruses.

☐ **What is the most common lethal, inheritable disease in the white population?**

Cystic fibrosis, an autosomal recessive disease occurring in 1/2,000 births.

☐ **Which class of antimicrobial agents are broadly effective against organisms that cause both typical and atypical pneumonias?**

Macrolides.

☐ **If a patient has a patchy infiltrate on a chest x-ray and bullous myringitis, what antibiotic should be prescribed?**

Macrolide antibiotics (Azithromycin, Erythromycin) for mycoplasma.

☐ **What secondary bacterial pneumonia often occurs following a viral pneumonia?**

Staphylococcus aureus pneumonia.

☐ **Should steroids be used in aspiration pneumonia?**

No.

☐ **What 3 findings should be present to consider a sputum sample adequate?**

1. > 25 PMNs;
2. < 10 squamous epithelial cells per low powered field;
3. a predominant bacterial organism.

PEDIATRIC EMERGENCY MEDICINE BOARD REVIEW

☐ **What is the indication for a chest tube in pneumothorax?**

Over 15% pneumothorax or a clinical indication, such as respiratory distress.

☐ **What therapy may increase the body's absorption of a pneumothorax or pneumomediastinum?**

A high inspired FiO2.

☐ **What is the profile of a classic patient with a spontaneous pneumothorax?**

Male, athletic, tall, slim, 15-35 years of age. Pneumothorax does not occur most frequently during exercise.

☐ **What accessory x-rays may be obtained for the diagnosis of pneumothorax?**

Expiratory film and lateral decubitus film on the affected side. CT scan of the chest.

☐ **Primary pulmonary hypertension is most common in what population?**

Young females. It is most often fatal within a few years.

☐ **What heart sounds are heard with pulmonary hypertension?**

A shortened second heart sound split and a louder P2.

☐ **T/F: Pulse oximetry is a reliable method for estimating oxyhemoglobin saturation in a patient suffering from CO poisoning.**

False. COHb has light absorbance that can lead to a falsely elevated pulse oximeter transduced saturation level. The calculated value from a standard ABG may also be falsely elevated. The oxygen saturation should be determined by using a co-oximeter that measures the amounts of unsaturated O2Hb, COHb, and metHb.

☐ **What is the half-life of COHb for a patient breathing room air? Breathing 100% oxygen? Breathing 100% oxygen at 3 atm of pressure?**

$FIO_2 = 0.21$	$FIO_2 = 1.0$	$FIO_2 = 1.0$ @ 3 atm
4–6 hour	40–90 minutes	20-30 minutes

☐ **A newborn is breathing rapidly and grunting. Intercostal retractions, nasal flaring, and cyanosis are noted. Auscultation shows decreased breath sounds and crackles. What is the diagnosis?**

Newborn respiratory distress syndrome, also known as hyaline membrane disease. X-rays show diffuse atelectasis. Treatment involves artificial surfactant and O2 administration through CPAP.

☐ **Which type of rhinitis is associated with anxious patients?**

Vasomotor rhinitis. This is a non-allergic rhinitis of unknown etiology that involves nasal vascular congestion.

PULMONARY

☐ **Which sinuses are fully formed in children?**

Only the ethmoid and the maxillary. Frontal sinuses form by age 10.

☐ **Differentiate between transudate and exudate.**

1. Transudate: Serum protein is < 0.5 and LDH is < 0.6. Most common with CHF, renal disease, and liver disease.
2. Exudate: Serum protein > 0.5, LDH > 0.6. Most common with infections, malignancy, and trauma.

☐ **Where do primary TB lesions most often occur in the lung?**

Primary lesions most often occur in the lower lobes of the lung. Reactivation occurs in the apices.

☐ **What are the side-effects of INH?**

Neuropathy, pyridoxine loss, lupus-like syndrome, and hepatitis.

☐ **What percentage of tuberculosis is drug-resistant?**

About 15%, but this mainly depends on the geographic location.

☐ **What are the classic signs and symptoms of TB?**

Night sweats, fever, weight loss, malaise, cough, and a greenish-yellow sputum most commonly seen in the mornings.

☐ **Where are some common extrapulmonary TB sites?**

The lymph node, bone, GI tract, GU tract, meninges, liver, and the pericardium.

☐ **A patient presents with cough, lethargy, dyspnea, conjunctivitis, glomerulonephritis, fever, and purulent sinusitis. What is the diagnosis?**

Wegener's granulomatosis. This is a necrotizing vasculitis and pulmonary granulomatosis that attacks the small arteries and veins. Treatment is with corticosteroids and cyclophosphamide.

☐ **Are sedatives beneficial for anxious asthmatic patients?**

Only if you are trying to kill them. However, sedatives are appropriate for use during Rapid Sequence Intubation (RSI).

☐ **What is the prognosis of childhood asthma?**

50% of all asthmatic children are symptom free within 10-20 years. In mild asthma, the remission rate is 50%, with only 5% experiencing severe disease. 95% of children with severe childhood asthma, however, become asthmatic adults.

☐ **Define extrinsic asthma.**

Asthmatic exacerbation following environmental exposure to allergens such as dust, pollens, and dander.

☐ **Define intrinsic asthma.**

Asthmatic exacerbation not associated with an increase in IgE or a positive skin.

☐ **When administering epinephrine in the treatment of asthma, how might the side effects of epinephrine be minimized?**

Side effects such as pallor, tremor, anxiety, palpitations, and headache can be minimized if doses of no more than 0.3 ml are given.

☐ **What are the historical risk factors for status asthmaticus?**

Chronic steroid-dependent asthma, prior ICU admission, prior intubation, recurrent ER visits in past 48 hours, sudden onset of severe respiratory compromise, poor therapy compliance, poor clinical recognition of attack severity, and hypoxic seizures.

☐ **What is the major side effect associated with the inhalation of N-acetylcysteine?**

Cough and bronchospasm, most likely due to irritation from the low pH (2.2) of the aerosol solution.

☐ **What is the effect of macrolide antibiotics (erythromycin, clarithromycin) on mucus hypersecretion?**

Some macrolide antibiotics have the ability to down regulate mucus secretion by an unknown mechanism. This is thought to be due to an anti-inflammatory activity.

☐ **What are the signs of a large obstructing foreign body in the larynx or trachea?**

Respiratory distress, stridor, inability to speak, cyanosis, loss of consciousness, death.

☐ **What are the symptoms of a smaller (distally lodged) foreign body?**

Cough, dyspnea, wheezing, chest pain, fever.

☐ **What is the procedure of choice for foreign body removal?**

Rigid bronchoscopy. Fiberoptic bronchoscopy is an alternate procedure in adults, not in children. If bronchoscopy fails, thoracotomy may be required.

☐ **What are the common radiographic findings in foreign body aspiration?**

Normal film, atelectasis, pneumonia, contralateral mediastinal shift (more marked during expiration), and visualization of the foreign body

PULMONARY

☐ **T/F: Aspiration of fluids with pH less than 5 produces chemical pneumonitis.**

False. Aspirate pH less than 2.5 produces chemical injury.

☐ **What is the antibiotic choice for gastric acid aspiration?**

None.

☐ **What is the role of corticosteroids in gastric acid aspiration?**

None.

☐ **What is the main priority in treating gastric acid aspiration?**

Maintenance of oxygenation. Intubation, ventilation, and PEEP (positive end expiratory pressure) may be required.

☐ **What are the radiographic manifestations of acid aspiration?**

Varied, may be bilateral diffuse infiltrates, irregular "patchy" infiltrates, or lobar infiltrates.

☐ **What is the usual source of infected aspirated material?**

The oropharynx.

☐ **What pleuropulmonary infections may occur after aspirating infected oropharyngeal material?**

Necrotizing pneumonia, lung abscess, "typical" pneumonia, empyema.

☐ **What is the predominant oropharyngeal flora in outpatients?**

Anaerobes. Community acquired aspiration is usually anaerobic. The most common aerobes involved are Streptococcus species.

☐ **What is the antibiotic of choice for outpatient acquired infectious aspiration pneumonia?**

Clindamycin, high dose penicillin or unasyn.

☐ **What is the bacteriology of patients with inpatient acquired infectious aspiration pneumonia?**

Mixed aerobic and anaerobic organisms. Unlike outpatients, Staphylococcus aureus, E. coli, Pseudomonas aeruginosa, and Proteus species are common.

☐ **What are the consequences of aspirating small (non-obstructing) food particles?**

Inflammation and hypoxemia, may result in chronic bronchiolitis or granulomatosis.

PEDIATRIC EMERGENCY MEDICINE BOARD REVIEW

☐ **Name the commonly aspirated hydrocarbons.**

Gasoline, kerosene, furniture polish, lighter fluid.

☐ **What are the effects of hydrocarbon aspiration?**

Hypoxemia, intrapulmonary shunting, pulmonary edema, hemoptysis, respiratory failure.

☐ **What is the role of emesis induction or gastric lavage in hydrocarbon ingestion?**

These maneuvers are not recommended as the patient may aspirate during regurgitation.

☐ **Lipoid pneumonia is associated with chronic aspiration of what?**

Mineral oil, animal or vegetable oils, oil-based nose drops.

☐ **What is the annual U.S. mortality from drowning?**

9000. There are 500 near drowning for each drowning.

☐ **What is the definition of drowning?**

Death due to suffocation by submersion in water.

☐ **Can drowning occur without aspiration of water?**

Yes. 10% of victims die from intense laryngospasm.

☐ **How much fluid can be drained from the lungs after drowning in fresh water?**

Little, as hypotonic fresh water is rapidly absorbed from the lung, unlike salt water which is hypertonic and is retained. There is no difference in outcome.

☐ **What is the priority in treatment of a submersion incident?**

Maintenance of oxygenation.

☐ **When is the Heimlich maneuver applied after a submersion/immersion incident?**

Never. This maneuver increases the chance of causing injury.

☐ **What are the indications for intubation and ventilation after a submersion/immersion incident?**

Apnea, pulselessness, altered mental status, severe hypoxemia, respiratory acidosis.

PULMONARY

☐ **Resistance to airflow in the airway is inversely proportional to what?**

The fourth power of the airway radius.

☐ **What is the most common etiology of croup or laryngotracheobronchitis?**

Parainfluenza virus.

☐ **What is the most common organism isolated from patients with bacterial tracheitis?**

Staphylococcus aureus.

☐ **How do retropharyngeal abscesses arise?**

Lymphatic spread of infections in the nasopharynx, oropharynx, or external auditory canal.

☐ **What is the most common type of tracheo-esophageal fistula?**

Blind esophageal pouch with a fistulous connection of the trachea to the distal esophagus.

☐ **The croup score is used to assess the degree of respiratory distress accompanying an illness. Which symptoms are evaluated in the score?**

Stridor, cough, air entry, flaring/retractions, and color.

☐ **Why does a newborn with bilateral choanal astresia present with severe respiratory distress that is relieved by the insertion of an oral airway?**

Because the newborn is an obligatory nose breather.

☐ **According to Poiseuille's law, if the radius of the conducting airway is reduced from 4 mm to 2 mm, by how much will the airflow resistance increase?**

Sixteen-fold.

☐ **The funnel-shaped narrowing of the glottis and subglottic airway, commonly known as steeple sign, is observed in what type of extrathoracic airway obstruction?**

Croup or laryngotracheobronchitis.

☐ **Name one of the conditions which makes a child a likely candidate for tonsillectomy.**

Recurrent tonsillar abscesses.

☐ **A six-year-old male complains of dysphagia and fever. Tonsillar exudates and anterior cervical adenitis are noted. What sign differentiates Group A strep from other causes of pharyngitis?**

A sandpaper-like rash.

PEDIATRIC EMERGENCY MEDICINE BOARD REVIEW

☐ A 4-year-old has a high fever, hoarseness, and increased stridor of 3 hours duration. He is ill appearing, with a temperature of 40o C (104o F), inspiratory stridor, drooling and intercostals retractions. He prefers to sit up. What is the most likely diagnosis?

Epiglottitis.

☐ Foreign body aspiration occurs most commonly in which age group?

1-3 years of age, with slight male predominance.

☐ Name the most common offender in foreign body aspiration.

Organic substances such as nuts and corn.

☐ What are the McConnochie criteria for the diagnosis of bronchiolitis?

1. Acute expiratory wheezing,
2. Age 6 months or less,
3. Signs of viral illness, such as fever or coryza,
4. Without pneumonia or atopy,
5. The first such episode,

☐ What will be the predominant symptom of a child who has aspirated a foreign body which is lodged in his/her trachea?

Stridor.

☐ What is the best single predictor of pneumonia in children?

Tachypnea.

☐ The predominant auscultatory finding in a child with a foreign body lodged in the right mainstem bronchus:

Expiratory wheezing.

☐ An 11-month old male presents with rhinorrhea, cough and increasing respiratory distress. Treatment with bronchodilators have produced equivocal results. He is now in obvious distress with retractions, grunting, prolonged expiratory time, inspiratory and expiratory wheezes. Chest x-rays shows a tracheal shift to the left and hyperexpansion of the right chest. What is the most likely diagnosis?

Foreign body aspiration.

☐ The total work of breathing is divided into two parts. Name them.

1. To overcome lung and chest wall compliance.
2. To overcome airway and tissue resistance.

PULMONARY

☐ **Total work of breathing in neonates and infants is lowest at what respiratory rate?**

35-40/minute.

☐ **Why are infants more susceptible to inflammatory changes in the airway?**

Because the airway in infants is smaller, and is also much more compliant.

☐ **Grunting is usually more prominent in what type of respiratory pathology?**

Typically in smaller airway disease such as bronchiolitis, or in diseases with loss of functional residual capacity such as pneumonia or pulmonary edema because grunting is an effort to maintain positive airway pressure during expiration.

☐ **What is Sampter's Triad?**

Asthma, nasal polyps, and aspirin allergy.

☐ **What are the criteria associated with severe asthma and likelihood of hospital admission?**

A pretreatment PEFR <25% predicted, pulsus paradoxus > 10 mm Hg on presentation, failure of PEFR to rise > 15% after treatment, or posttreatment PEFR <60% predicted or SaO2 <93%.

☐ **What is the dose of epinephrine in acute asthma?**

0.01 mg/kg SC of 1:1000, repeated every 15-20 minutes as meeded. Max dose of 0.3 mg.

☐ **What is the dose of nebulized albuterol?**

0.1-0.15 mg/kg of a 2 mg/5 ml solution.

☐ **What are the high yield indications for chest radiography in suspected pneumonia?**

Tachypnea, rales, decreased breath sounds, stridor, retractions, or cyanosis.

GASTROINTESTINAL

☐ **What is the most common cause of pain of anal origin, in a child?**

Anal Fissure. This is due to a tear in the anal epithelium distal to the dentate line.

☐ **What are some common conditions that mimic acute appendicitis?**

Mesenteric lymphadenitis, PID, Mittlesmertz, gastroenteritis, ovarian cyst, ovarian torsion, kidney stone, Meckel's diverticulum and Crohn's disease.

☐ **An 11 month old child presents with a small, hard, tender perianal mass. Vital signs are normal and the infant does not appear toxic. What is the diagnosis and treatment?**

Perianal abscess. Oral antibiotics and no surgery in children < 1 year of age with a small abscess.

☐ **What is the most common cause of an anal fistula (fistula-in-ano)?**

Perianal abscess. May develop spontaneously or after surgical drainage. Most resolve without intervention.

☐ **What conditions are associated with an atypical presentation of acute appendicitis?**

Situs inversus viscerum, malrotation, hypermobile cecum, long pelvic appendix, and pregnancy (1/2200).

☐ **What are the most frequent symptoms of acute appendicitis?**

Anorexia and pain. The classical presentations of anorexia and periumbilical pain with progression to constant RLQ pain are present in only 95% of the cases.

☐ **A 4 year old female presents with right lower quandrant pain, vomiting and a normal WBC. The consulting surgeon states that an acute appendicitis has been ruled out. Is he correct?**

No. 15% to 20% of cases will have a normal WBC and up to 80% will have a normal count in the first 24 hours of symptoms.

☐ **What intraabdominal pathology should be assumed in a pregnant female with right upper quadrant pain until proven otherwise?**

Acute appendicitis.

PEDIATRIC EMERGENCY MEDICINE BOARD REVIEW

☐ **An 8 year old male (weight 25kg or 55 lbs) complains of right lower quadrant pain. You are concerned about an acute appendicitis. What is the initial test of choice?**

Ultrasound of the RLQ. If done by an experienced sonographer the sensitivity is 75% - 90% and the specificity is 85% - 100%.

☐ **What are the ultrasound findings in an acute appendicitis?**

Noncompressible, blind-ending tubular structure in the longitudinal axis, measuring > 6 cm in diameter. Target appearance (bagal sign) on the transverse view. Free fluid around the appendix. Appendicolith. Appendiceal wall hyperemia

☐ **Which method is more sensitive for locating the source of GI bleeding, a radioactive Tc-labeled red cell scan or an angiography?**

A bleeding scan can detect a site bleeding at a rate as low as 0.12 mL/minute, while angiography requires rapid bleeding (i.e., greater than 0.5 mL/min). The advantages of angiography are diagnosis along with possible treatment (embolization, intra-arterial vasoconstrictors).

☐ **A child with gastroesophageal reflux develops melena and vomits a large amount of bright red blood. Is esophagitis the likely cause?**

No. Capillary bleeding rarely causes impressive acute blood loss. Arterial bleeding from a complicated ulcer, variceal bleeding, or A-V malformation are more probable.

☐ **What is the most common cause of upper GI bleeding in children 1 month to 1 year of age?**

Esophagitis caused by gastroesophageal reflux.

☐ **Repeated violent bouts of vomiting can result in both Mallory-Weiss tears and Boerhaave's syndrome. Differentiate between the two.**

1. Mallory-Weiss tears involve the submucosa and mucosa, typically in the right posterolateral wall of the GE junction.
2. Boerhaave syndrome is a full-thickness tear, usually in the unsupported left posterolateral wall of the abdominal esophagus.

☐ **A 6 month old infant is constipated, flaccid, and only stares straight ahead. He is afebrile and won't take a bottle, even when honey is placed on the tip. What is the likely diagnosis?**

Infantile botulism. Infants ingest spores that are in the environment, most commonly in honey. The spores then grow and produce toxins within the host.

☐ **How does infantile botulism differ from food-borne botulism?**

Infantile botulism is caused by the spores of C. botulinum while adult food-borne botulism arises from the ingestion of C. botulinum neurotoxins.

GASTROINTESTINAL

☐ **20 minutes after feeding, a 3 month old presents with an episode of opisthotonic posturing accompanied by a guttural cry. An extensive neurologic, metabolic and cardiopulmonary work up are negative. What is the likely condition?**

Sandifer's syndrome. A benign condition associated with gastroesophageal reflux, esophagitis, or the presence of a hiatal hernia. It is not life threatening and most children outgrow the spasm in later childhood.

☐ **What is the most common cause of rectal bleeding in infants?**

Anal Fissure.

☐ **What findings indicate that a child's gastroesophageal reflux disease is benign?**

Nonbilious vomiting that begins shortly after birth, no sudden starting or ending point, nonprojectile vomiting and good weight gain.

☐ **What are the risk factors for the development of acalculous cholecystitis in a child?**

Shock, sepsis, hyperalimentation, prolonged fasting, IV narcotics and multiple transfusions.

☐ **A patient with a history of gallstones presents with an acute, postprandial right upper quadrant pain. What's the KUB likely to show?**

Nothing specific. Only around 10% of gallstones are radiopaque. Complications of cholelithiasis, such as emphysematous cholecystitis, perforation, and pneumobilia are uncommon but useful radiographic findings when present.

☐ **A child with sickle-cell disease presents with fever with shaking chills, RUQ pain, and jaundice. What is the diagnosis?**

Charcot's triad suggests ascending cholangitis. The precipitating cause is probably pigment stones resulting from chronic hemolysis.

☐ **A 12-year-old male complains of 2 days of rice water stools, muscle cramps, and extreme fatigue. He looks pale, dehydrated, and very ill. The patient states that he has just returned from India. What is the diagnosis?**

Cholera. It usually develops in travelers returning from endemic areas, such as India, Africa, southeast Asia, Central and South America, and the Middle East. It can be prevented if the drinking water is purified and raw, unpeeled fruits/vegetables and seafood are avoided.

☐ **What are the most common ocular abnormalities seen in patients with Crohn disease?**

Episcleritis and anterior uveitis.

☐ **What are the most common cutaneous manifestations of Crohn disease?**

Erythema nodosum and pyoderma gangrenosum (ulcerative skin condition).

PEDIATRIC EMERGENCY MEDICINE BOARD REVIEW

☐ **What is the most common complication of gallstones in children?**

Pancreatitis.

☐ **What are the common bacteria seen in acute cholecystitis?**

E coli, enterococci, and Klebsiella species.

☐ **A 9 year old male with Kawasaki disease, presents with RUQ pain, jaundice, hepatomegaly, elevated LFTs, and an enlarged gallbladder without stones on ultrasound. What is the diagnosis?**

Gallbladder hydrops (gallbladder mucocele). Also associated with mesenteric adenitis, typhoid fever, streptococcal pharyngitis, hepatitis, nephrotic syndrome, and leptospirosis.

☐ **What stool studies are crucial for evaluating acute diarrhea?**

Acute diarrhea is common and self-limiting therefore, testing is rarely required. In sick patients or those at risk for complications, (recently hospitalized, recent antibiotic use ,or immunocompromised), consider a stool culture, smear for WBCs and RBCs, Gram's stains, ova and parasites, roto virus test and c-difficile toxin.

☐ **A 15 year old female presents with chronic intermittent abdominal pain, diarrhea, anorexia, weight loss and growth deceleration. Her Tanner stage indicates pubertal delay and she has perianal skin tags. What disorder are you most concerned about?**

Crohn disease. Children may manefest growth and pubertal delay long before the GI symptoms start. Perinal disease (perinanal tags, abscesses, fistulae, fissures) are present in 45% of patients.

☐ **A 6-month old male presents with a fever, vomiting, watery diarrhea and signs of dehydration. What is the most likely inciting pathogen?**

Rotavirus. This is the most common cause of infectious diarrhea in children < 5 years of age. Worldwide it is responsible for 120 million cases of infantile diarrhea and 500,000 deaths annually.

☐ **What is the most common cause of bacterial diarrhea?**

Campylobacter jejuni (6%-8%). Salmonella (7%), E Coli (2%-5%), Shigella (1%-3%).

☐ **What is the most common foodborne pathogen causing vomiting and diarrhea in the United States?**

Norovirus. It was previously called Norwalk virus until it was found to be a species in the genus Norovirus.

☐ **What are the risk factors for cholelithiasis in infancy.**

Hyperalimentation, abdominal surgery, sepsis, bronchopulmonary dysplasia, hemolytic disease, necrotizing enterocolitis, and hepatobiliary disease.

GASTROINTESTINAL

☐ **What is the most common type of gallstone?**

Cholesterol stones (>50%). Pigment stones are black, radiopaque, and composed of calcium bilirubinate. Brown stones are composed of calcium salts of bilirubin, stearic acid, lecithin, and palmitic acid and are associated with infection.

☐ **What is the most common cause of acute and chronic cholecystitis?**

Gallstones.

☐ **A 3 week old patient presents with fever, worsening jaundice, and elevated transaminases soon after surgical correction of biliary atresia. What complication are you most concerned about?**

Acute cholangitis.

☐ **An 8 month old male presents with jaundice, hepatomegaly, acholic stools and an elevated lipase. A RUQ ultrasound shows a large cystic structure without stones lying next to the liver. What is the likely diagnosis?**

Choledochal cyst. This is a cystic dilatation of the common bile duct. Surgical excision of the cyst is required.

☐ **What substances can color both the vomitus and stool red?**

Red food dye, cefdinir, rifampin, beets and blueberries.

☐ **Vomiting that develops within 6 hours of a meal is most probably caused by what?**

An ingested of a pre-formed enterotoxin. Staphylococcal enterotoxin A and B are the most common ones.

☐ **What is the most cause of traveler's diarrhea?**

Enterotoxigenic E coli. Other bacteria include Campylobacter jejuni, Shigella species, and Salmonella species. Starts about 3-7 days after arrival in a foreign land.

☐ **What is the treatment for traveler's diarrhea?**

To adequately cover for E. Coli, the prefered treatment is ciprofloxin with trimethoprim/sulfamethoxazole (Bactrim, Septra) as an alternative.

☐ **What is the definition of acute diarrhea?**

3 loose stools per day for < 14 days. Chronic diarrhea is when it lasts > 14 days.

☐ **What is the most common esophageal disorder in all age groups?**

Gastroesophageal reflux. Seen in 65% of healthy children by 5 months of age.

PEDIATRIC EMERGENCY MEDICINE BOARD REVIEW

☐ **What are the contributing factors in gastroesophageal disease?**

Immaturity of the esophagus and lower esophageal sphincter, short length of the abdominal esophagus and a liquid diet.

☐ **What are the initial measures used to treat gastroesophageal reflux?**

Smaller frequent feeds and "burping," upright position for 45 minutes after feeding, and thickening of the formula with cereal. If medications are required, H2 blockers (ranitidine, famotidine) are the drugs of choice.

☐ **What are the ENT complications of gastroesophageal reflux?**

Dental erosions, otitis media, sinusitis, and laryngitis.

☐ **What are the most common entities aspirated in children?**

Food items such a nuts, sunflower seeds, poorly chewed meets and hotdogs, raisins, candies, grapes. Most foreign body aspirations occur in children < 3 years old.

☐ **What size objects rarely pass through the stomach and require endoscopic retrieval?**

Objects longer than 5 cm and wider than 2 cm.

☐ **Radiographs should be performed on all patients suspected of swallowing coins to determine the presence and location of the foreign body. How will the coin appear on a x-ray in the AP view?**

Coins in the esophagus lie in the frontal plane. Coins in the trachea lie in the sagittal plane.

☐ **What is the management of button batteries that have passed the esophagus?**

In the asymptomatic patient, repeat radiographs. The symptomatic patient and patients in whom the battery has not passed the pylorus after 48 hours require an endoscopic retrieval.

☐ **What history and physical exam findings make you concerned for an occult foreign body aspiration?**

A sudden episode of coughing or choking while eating along with esophageal pain, and abnormal airway sounds (wheezing, stridor, decreased breath sounds).

☐ **Which type of hepatitis is usually contracted through blood transfusion?**

Hepatitis C. Eighty five percent of post transfusion hepatitis is caused by hepatitis C.

☐ **The ingestion of which types of food can cause a false positive guaiac test?**

Red meat, peroxidase-containing fruits, cauliflower, broccoli, turnips, or radishes.

GASTROINTESTINAL

☐ **What are the most common pathogens causing bloody diarrhea?**

Salmonella, Shigella, Campylobacter jejuni, Yersinia enterocolitica, E coli, Clostridium difficile, and Entamoeba histolytica.

☐ **An ill appearing, afebrile 1 month old female presents with a sudden onset of bilious vomiting and severe unremitting pain. What is the most likely diagnosis?**

Midgut volvulus. Bilious vomiting is the most common and important finding (75% of cases).

☐ **An 8 year old male presents with abdominal pain, vomiting, and abdominal distention. Is a midgut volvulus high on the differential?**

No. Even though anything can happen at anytime, 40% of cases are diagnosed in the 1st week, 75% by 1 year, and 99% by 2 years of age.

☐ **What does the phrase "bilious vomiting" mean?**

Yellow or greenish colored emesis. This indicates the presences of bile and results from a severe obstruction.

☐ **What are the radiographic findings on a KUB in a patient with a midgut volvulus?**

Double bubble sign (air-fluid levels in a dilated stomach and duodenum) and multiple loops of dialted bowel.

☐ **What is the test of choice to diagnose a midgut volvulus?**

Upper GI series. The duodenal c-loop crosses to the right of the midline and contrast dye abruptly ends.

☐ **What are the most common bacterial causing diarrhea in children going to daycare?**

Campylobcter and Shigella.

☐ **A 15 year old female with a past history of intermittent crampy abdominal pain with bloody stools, presents with fever, abdominal pain, and signs of dehydration. On exam you note a distended abdomen and an altered level of consciousness. What condition are you concerned about?**

Toxic Megacolon. Most children do not have the hallmark findings of altered level of consciousness and hypotension.

☐ **What is the most common cause of toxic megacolon in children (past the neonatal period).?**

Complication from ulcerative colitis. The next most common cause is clostridium difficile. This is a newer phenomena due to the increase incidence of this pathogen. Toxic megacolon is believed to be the result of the inflammation causing a decrease in the smooth muscle contractility.

PEDIATRIC EMERGENCY MEDICINE BOARD REVIEW

☐ **What are the most common parasitic causes of diarrhea?**

Cryptosporidium and Giardia lamblia.

☐ **Explain the different hepatitis B serologic test and what there presence indicates.**

- Hepatitis B surface antigen (HBsAG): Detected in high levels during acute or chronic hepatitis B infection. Its presence indicates that the person is infectious.
- Hepatitis B surface antibody (anti-HBs): Indicates recovery and immunity from hepatitis B infection. Anti-HBs develops in persons who have been successfully vaccinated.
- Hepatitis B core antibody (anti-HBc): Its presence indicates previous or ongoing infection with hepatitis B in an undefined time frame.
- IgM antibody to hepatitis B (IgM anti-HBc): Its presence indicates recent hepatits B infection (< 6 months)

☐ **What lab tests are the best indicator of continued hepatocellular damage in a patient with hepatitis?**

The serum transaminases (ALT, AST).

☐ **What is the most common identifiable cause of fulminant hepatic failure in children?**

Viral hepatitis in 50% of cases and drug-induced hepatotoxicty (acetaminophen, salicylates, methanol, isoniazid, valproic acid) in 25% of cases.

☐ **What is the most common metabolic cause of fulminant hepatic failure in children?**

Wilson disease. Children may not present with jaundice and usually have nonspecific complaints.

☐ **What findings indicate a poor prognosis in patients with fulminant hepatic failure?**

Jaundice of > 7 days, PT > 50 seconds, and Serum bilirubin > 17.5.

☐ **What is the most common cause of diarrhea in daycare children?**

Rotavirus. The incubation period is 1-7 days with a duration of 4-8 days. It most commonly occurs between December and May.

☐ **A 3 month old presents with an umbilical hernia. The parent is worried and wants it repaired. What do you tell her?**

These are very common in children, rarely incarcerate, and spontaneously close by 2 to 5 years of age.

☐ **What is the most common type of hydrocele?**

Communicating type. This results when the proximal portion of the process vaginalis remains patent, allowing fluid from the abdominal cavity to freely enter the scrotal sac. A noncommunicating type occurs when closure is present proximal but fluid remains trapped within the distal tunica vaginalis.

GASTROINTESTINAL

☐ **What is the most common neonatal gastrointestinal emergency.**

Necrotizing enterocolitis (NEC). This condition primarily affect low-birth-weight, premature infants. Who received an umbilical catheter, and aggressive hypertonic enteral feedings.

☐ **A 4 week old male with a past history of NEC presents with abdominal distention, poor feeding, and appears toxic. What is the most likely GI cause?**

Intestinal obstruction. Necrotizing enterocolitis (NEC) can cause intestinal strictures most commonly in the terminal ileum and colon..

☐ **Distinguish between a groin hernia, a hydrocele, and a lymph node.**

Hydroceles transilluminate and are non-tender. Lymph nodes are freely moving and firm. Hernias don't transilluminate and may produce bowel sounds.

☐ **A 14 year old male tells you that he has had a "bulge" in his groin for 2 days. The pain has become progressively worse and he can't stop vomiting. You palpate a tender mass in the groin. What is the diagnosis and treatment?**

Incarcerated (and possible strangulated) inguinal hernia. The treatment is IV fluids, pain control, and immediate surgical consultation. It is best not to reduce a long-standing, tender, incarcerated hernia for fear of putting dead bowel back into the abdomen.

☐ **A 3 week old female presents with cough, fever, vomiting, and poor feeding. A chest x-rays reveals the presence of the gastric bubble in the thorax. What condition does this indicate and how could you better diagnose the problem.**

Congenital diaphragmatic hernia. A nasogastric tube can be inserted and dye given for better delineation of the anatomy. 90% occur on the left side through the foramen of Bochdalek.

☐ **What is the most common cause of esophageal obstruction in neonates?**

Tracheoesophageal fistula. The most common type of TEF (90% of cases) results in the proximal esophagus terminating in a blind pouch, while the distal esophagus communicates with the trachea.

☐ **What simple test can distinguish between conjugated and unconjugated hyperbilirubinemia?**

A dipstick test for urobilinogen, which reflects conjugated (water soluble) hyperbilirubinemia.

☐ **In addition to conjugated hyperbilirubinemia, what liver function abnormalities suggest biliary tract disease?**

Elevated alkaline phosphatase out of proportion to transaminases.

☐ **What is the most common condition requiring medical attention in newborns?**

Jaundice. Most commonly it is unconjugated hyperbilirubinemia (indirect) due to physiologic jaundice, breast milk, hemolysis, and bruising from birth trauma.

☐ **Why is it normal for neonates to have an elevated unconjugated bilirubin?**

Neonates have an increased breakdown of fetal RBCs due to their half-life, and higher erythrocyte mass. Hepatic excretory capacity is low due to a lower level of circulating binding protein and a decreased enzyme activity. This leads to a decrease in conjugated bilirubin (direct) and an increase in the unconjugated (indirect) form which is harder to excrete.

☐ **A newborn develops jaudice within the first 24 hours of life. He is otherwise healthy appearing and mom is ready to take him home. What should be your response?**

Jaundice that is visible during the first 24 hours is likely nonphysiologic and always requires further workup.

☐ **What is the differential diagnosis for the patient describe in the previous question?**

Hemolysis (ABO, Rh incompatibility), congenital infection (TORCH), excessive bruising from birth (cephalhematoma), and aquired infection.

☐ **When does physiologic jaundice of infancy peak?**

3rd to the 5th day of life. Physiologic jaundice is icterus that is not pathologic.

☐ **A neonate presents with detectible jaundice to the face and eyes. What is the estimated serum bilirubin level?**

7-8 mg/dL. Shoulder/torso 8-10 mg/dL, lower body 10-12 mg/dL, Generalized > 12 mg/dL.

☐ **What is the most common cause of jaundice in a newborn from day 2-3?**

Physiologic jaundice.

☐ **Differentiate between breast feeding jaundice and breast milk jaundice?**

Breast feeding jaundice occurs in the first 3 days of life and is caused by insufficient production or intake of breast milk leading to dehydration. Breast milk jaundice develops 4-7 days of life, persists longer than physiologic jaundice, and has no identifiable cause.

☐ **A 10 day old neonate who was delivered at home, is brought to the ED with jaundice. The patient presents with a tremor, extensor rigidity, loss of suck reflex, lethargy, and seizures. What condition are you concerned about?**

Kernicterus due to an elevated bilirubin. The potential causes are acquired infection, biliary atresia, hepatitis, spherocytosis, G6PD deficiency, hemolysis from drugs, and hypothyroidism.

☐ **What is the treatment for the patient in the previous question?**

The definitive treatment is exchange transfusion. A full septic and metabolic work up and phototherapy are also indicated but this should not delay the exchange transfusion.

GASTROINTESTINAL

☐ **A 2 week old presents with a complaint by the parent of "he is just not acting right and he is yellow." Lab work reveals a significantly elevated direct (conjugated) bilirubin. What are the potential causes.**

Neonatal direct (conjugated) hyperbilirubinemia is never normal. The most common causes include biliary atresia, extrahepatic biliary obstruction, neonatal hepatitis, and alpha 1-antitrypsin deficiency.

☐ **What is the most common cause of unconjugated (indirect) hyperbilirubinemia beyond the neonatal period?**

Hemolysis due to sickle cell disease, hereditary spherocytosis, and G6PD deficiency.

☐ **What is the most common cause of conjugated (direct) hyperbilirubinemia beyond the neonatal period?**

Biliary obstruction or hepatocellular injury.

☐ **At what bilirubin level do you start phototherapy?**

20 mg/dL.

☐ **What is the most common cause of bowel obstruction in children between 3 months and 6 years of age?**

Intussusception. Ileo-colic region is most affected. 60% of diagnosed patients are < 1 yr. 80% of cases occur before 2 years of age. Most common cause is idiopathic but occurs in peaks in spring and autumn possible correlating with adenovirus infection.

☐ **What is the treatment for intussusception?**

An air enema is both a diagnostic tool and curative (75-90% successful), by reducing the intussusception. There is a 10% recurrance rate with most cases returning within 72 hours. Surgery is required for refractory cases.

☐ **A 16 month old female presents with intermittant, colicky abdominal pain, vomiting, and bloody stools. What is the most likely diagnosis?**

Intussusception. This is the classic triad for but is only seen in < 20% of cases. Lethargy is a common finding and may be the only finding in a patient with this condition.

☐ **What are the radiographic findings on a KUB in a patient with intussusception?**

-Pseudokidney or target sign: Mass in the RUQ that has the shape of a kidney.
-Crescent sign: Intussusceptum lead point protruding into a gas filled pocket.
-Obscured liver edge.

☐ **What are the ultrasound findings in a patient with intussusception?**

Coiled spring sign, sleeve sign, target sign. US has a > 95% sensitivity and > 95% specificity.

PEDIATRIC EMERGENCY MEDICINE BOARD REVIEW

☐ **A six year old male presents with severe abdominal pain with rebound and guarding on exam. You obtain an abdominal x-ray series to rule out the presences of free air. If negative, have you ruled out a ruptured hollow viscus?**

No. Approximately 30% of patients with have an initial negative x-ray looking for free air. To better your yield, keep the patient in either the upright or left lateral decubitus position for at least 10 minutes prior to performing x-rays.

☐ **Name two endocrine problems that cause peptic ulcers.**

Zollinger-Ellison syndrome and hyperparathyroidism (hypercalcemia).

☐ **A 2 year old male presents with rectal bleeding. The child is happy, nontoxic and vital signs are normal. What is the most likely diagnosis?**

Meckel's diverticulum. AVM is also a consideration but less common. Remember the rule of 2s: 2% of the population, 2:1 male/female ration, 2 feet from the ileocecal junction, 2 inches in length, 2 cm in diameter, and asymptomatic before age 2.

☐ **What is a Meckel's diverticulum and what is the big deal about them?**

Remnant of the failed obliteration of the omphalomesenteric duct which usually occurs during the 5th week of embryologic development. Problems can occur when ectopic tissue (gastric in 80% of cases) is present.

☐ **What is the test of choice in diagnosing a Meckel's diverticulum?**

Meckel Scan. Technetium-99m pertechnetate is injected and taken up by gastric mucosal cells. Sensitivity is between 80-90%.

☐ **A 4 year old male presents with a painless abdominal mass, hematuria, and an elevated blood pressure. What is the most likely diagnosis?**

Wilm's tumor. This is the most common abdominal and renal malignancy in childhood.

☐ **Do pilonidal abscesses communicate with the anal canal?**

No. They are virtually always midline and overlie the lower sacrum. Posterior-opening, horseshoe-type anorectal fistulas can find their way to the lower sacrum, but are rarely in the midline. Remember Goodsall's rule.

☐ **What is the most common cause of gastric obstruction in infants?**

Pyloric stenosis. 1:250 births with a male to female ratio of 4:1. Most cases are seen between 2-5 weeks after birth and rarely seen past 8 weeks of age.

GASTROINTESTINAL

☐ **A 6 month old male presents with continuous crying associated with bilious vomiting. The patient stops while in the ER and is sent home with a diagnosis of colic. Is this the correct diagnosis?**

No. Bilious vomiting in an infant is always considered pathologic and requires a work up. Colic follows the rule of 3's: 3 weeks of age, 3 hours of crying per day, 3 days per week, for 3 weeks.

☐ **A 3 week old males presents with nonbilious vomiting 10-20 minutes after each feeding. The child appears nontoxic with normal vital signs. Lab test show hypokalemia, an elevated CO_2, and a decrease chloride. What is the most likely diagnosis?**

Pyloric stenosis. Vomiting causes a decrease in the potassium, excessive loss of HCl, with a resultant hypochloremic metabolic alkalosis. Patients may appear normal with non-projectile vomiting on initial presentation.

☐ **What substance can mimic melena?**

Iron supplements, dark chocolate, spinach, cranberries, grape juice, bismuth, and blueberries.

☐ **A 10-year-old child presents with a three day history of episodic abdominal pain and hematochezia. Examination reveals a palpable purpuric rash mostly on the lower limbs and the latelet count is normal. What is the most likely cause?**

Henoch-Schonlein purpura (anaphylactoid purpura).

☐ **What is the test of choice in diagnosing pyloric stenosis?**

Ultrasound. It approaches 100% sensitivity and specificity in experienced hands. The pyloris is > 4mm thick, 16mm long, and a "target sign" is present.

☐ **What are the three most common causes of upper gastrointestinal bleeding in neonates?**

Swallowed maternal blood, gastritis, and esophagitis.

☐ **What are the radiographic findings in a pediatric patient with toxic megacolon?**

A colon diameter of > 6cm in older children and > 4cm in children under 11 years of age. The ascending and transverse colon are most commonly affected with air-fluid levels present, and lose of visable haustra.

☐ **A 10 month old female presents with lethargy. Mom states she has had multiple crying episodes over the past 2 days and blood was present in her diaper. Her abdominal exam is normal and. What GI problem are you concerned about.**

Intussusception. In 10% of cases lethargy is the only presenting sign.

☐ **What is the treatment for a patient with toxic megacolon?**

IV fluids, IV antibiotics, NG tube, IV hydrocortisone (if due to ulcerative colitis), and surgical consultation.

PEDIATRIC EMERGENCY MEDICINE BOARD REVIEW

☐ **Name the three areas of physiologic narrowing in the esophagus, where an ingested coin is likely to lodge.**

Below the cricopharyngeal muscle, at the level of the aortic arch, and just below the diaphragm.

☐ **What is the most common cause of constipation in all age groups?**

Functional and dietary.

☐ **A newborn (neonate) presents to the ED 3 days after birth. Parents state he has had no bowel movement since birth. What is the differential diagnosis?**

Imperforate anus, anal stenosis, meconium plug, meconium ileus, Hirschsprung disease, volvulus and hypothyroidism.

☐ **The parents of a 3 week old male, present to the ED concerned because there child is having 5 bowel movements a day. He is breast-fed and the stools are well formed. The physical exam and vital signs are normal. What is your advice?**

Newborns have 3-4 bowel movements a day with breast-fed newborns having more. The number of stools per day decrease over time. By age 3, children have on average 1 per day. There is variation with many children having 1 stool every 3-4 days.

☐ **A 4 week old female presents with green-colored emesis after each feeding. The child appears irritable with a firm distended abdomen. What disorder are you concerned about and what is your treatment plan?**

Malrotation with a midgut volvulus. Treatment includes, IV rehydration, NG to suction, and consult a surgeon. 40% of cases occur in the first week of life, with 75% occurring by 1 year.

☐ **What is the most common mechanical reason for malrotation?**

Failure of the cecum to move into the right lower quadrant, so that the bands fixing it to the posterior abdominal wall cross over and may obstruct the duodenum.

☐ **Do infants with a midgut volvulus, always present with bilious vomiting?**

No. About 25% of infants under 2 months do not but bilious vomiting in an infant is serious until proven otherwise.

☐ **What is the diagnostic test of choice in diagnosing a midgut volvulus?**

Upper GI. Signs include, contrast abruptly tapering off and the duodenal c-loop to the right of the spine. Plain films are of little value but may show a double-bubble sign and multiple loops of dilated bowel.

☐ **What is the most common known cause of pancreatitis in children?**

Trauma (25%). Pancreaticobiliary disease (15%), Multisystem disease (14%), drugs and toxins (12%), viral infections (10%), hereditary/metabolic disorders (2%). Etiology is unknown in 25% of cases.

GASTROINTESTINAL

☐ **What is the most common finding in children with acute pancreatitis?**

1) Abdominal pain, 2) abdominal tenderness, 3) vomiting.

☐ **What are the most common presenting signs and symptoms of ulcerative colitis?**

Rectal bleeding, diarrhea and abdominal pain. As the disease progresses you find anorexia, mild anemia, weight loss and growth delays. It is usually diagnosed between ages 5 and 16.

☐ **What is the most common congenital malformation of the gastrointestinal tract?**

Meckel's diverticulum.

☐ **Name the most common non-infectious causes of chronic diarrhea in children.**

Food allergy, food intolerance, chronic nonspecific diarrhea (CNSD), lactase deficiency, irritable colon, encopresis, metabolic/malabsorption disease, ulcerative colitis, regional enteritis, and Hirschsprung's disease.

☐ **What is the most common reason for "spitting up" in young infant?**

Overfeeding! Parents frequently report that the child is vomiting.

☐ **A two week old male infant presents with persistent emesis and dehydration. Blood electrolytes reveal significant hyperkalemia. What potentially lesion must be ruled out?**

Congenital adrenal hyperplasia (CAH).

☐ **What is a common G.I. complication of nephrotic syndrome?**

Spontaneous bacterial peritonitis.

☐ **What are some lesions that can present as a mass protruding the rectum?**

Rectal polyp, rectal prolapse, and procidentia (prolapse of mucosa only), abscess.

☐ **What is the span of the liver at birth?**

4.5 - 5 cm.

☐ **What is the treatment for a prolapsed rectum?**

Simple manual reduction.

☐ **What is the most common cause of rectal prolapse?**

Excessive straining due to constipation or diarrhea.

PEDIATRIC EMERGENCY MEDICINE BOARD REVIEW

☐ **What is the most common cause of GI bleeding from an obstruction?**

Intussusception.

☐ **What age group is Reye's syndrome seen most commonly?**

4 - 12 years.

☐ **What is the major lethal factor in children with Reye's syndrome?**

Increased intracranial pressure, secondary to cerebral edema.

☐ **Name some medications associated with hepatic injury in children?**

Acetaminophen, Sulfonamide, Erythromycin, Ceftriaxone, INH, Valproic acid, Phenytoin.

☐ **What common NICU procedure can result in portal hypertension?**

Umbilical catheterization.

☐ **What is the most common presentation of childhood portal hypertension?**

Bleeding from esophageal varices.

☐ **What is the major indication for liver transplantation in the pediatric age group?**

Biliary atresia.

☐ **What metabolic condition manifests with jaundice, hypoglycemia, convulsions, cataracts, and mental retardation?**

Galactosemia.

☐ **Do the levels of liver-dependent coagulation factors decrease after birth?**

Yes, they fall 48 - 72 hours after birth, and return to birth levels only by 7 to 10 days of life.

☐ **A healthy three-week old female child is brought to the emergency room with complaints of spitting up blood. She was a full-term birth, who has been thriving well on breast milk, and has gained 15oz since birth. CBC and liver enzymes are normal, but the prothrombin time, and partial thromboplastin time are prolonged. What would you do next?**

Administer 1 to 5 mg of Vit K IV.

☐ **What is the most commonly occurring malignant liver neoplasm in the pediatric age group?**

Hepatoblastoma.

GASTROINTESTINAL

☐ **6-12 hours after ingesting a wild mushroom, a child developed vomiting, diarrhea, and abdominal pain which has lasted for the past 3 days. Now the child is noted to be jaundiced. What do you suspect?**

Acute liver toxicity from Amanita mushroom poisoning.

☐ **An 11-year-old girl with cystic fibrosis presents with a one-day history of fever. On examination she has normal breath sounds and right upper abdominal quadrant pain. Labs reveal a leukocytosis, with an increase in serum bilirubin. What do you suspect?**

Acute calculous cholecystitis, seen in up to 4% of patients with cystic fibrosis.

☐ **A neonate with neonatal hemochromatosis is awaiting liver transplantation. He is presently being managed with deferoxamine. His anxious mother calls you up, and informs you that her son's diapers are stained brown. What is your diagnosis?**

Deferoxamine chelates iron, which is then excreted in the urine. When such urine stands for some time, it turns brown in color.

☐ **What is the incidence of meconium ileus in newborn infants with cystic fibrosis?**

Meconium ileus is most commonly seen in infants with cystic fibrosis, but less than 10% of infants with cystic fibrosis have meconium ileus.

☐ **What are the typical radiographic findings in meconium ileus?**

Lack of air in the distal colon; distended loops of small bowel associated with bubbles of stool and air-described as a ground-glass appearance or Neuhauser's sign.

☐ **What is Charcot's triad?**

The triad of right upper quadrant abdominal pain, fever, and jaundice. This is seen in cholangitis.

☐ **Which infectious agents are commonly associated with necrotizing enterocolitis?**

Gram negative enteric bacilli, E. coli, Klebsiella pneumonia, and Enterococcus.

☐ **Which microbial agents are associated with mesenteric adenitis and pseudoappendicitis?**

Yersinia.

☐ **Which two viral illnesses are prodromes for Reye syndrome?**

Varicella (chicken pox) and influenzae B.

☐ **What are the signs and symptoms of Reye syndrome.**

Irritability, combativeness, and lethargy, right upper quadrant tenderness, history of influenzae B or recent chicken pox, papilledema, hypoglycemia, and seizures. Lab results reveal hypoglycemia, an ammonia level greater than 20 times normal, and a bilirubin level that is normal.

PEDIATRIC EMERGENCY MEDICINE BOARD REVIEW

☐ **What is the most common cause of intussusception?**

95% are idiopathic, usually from mucosal edema and lymphoid hyperplasia following gastroenteritis. The rest are a result of a lead point, most commonly a Meckel diverticulum.

☐ **What is the prognosis if air is discovered in the portal venous system?**

Portal venous air is associated with bowel necrosis and impending perforation. While portal venous air suggests a poor prognosis, with aggressive treatment, air may be transient and does not imply a hopeless outcome.

☐ **What is a sentinel loop?**

A distended loop of bowel seen on x-ray that lies near a localized inflammatory process. This is a clue to an underlying inflammatory process adjacent to the distended bowel. It can often be seen in pancreatitis or appendicitis.

☐ **A 1-day-old with choking has an orogastric tube placed. X-ray reveals the tube is coiled in the proximal esophagus. Air is present in the stomach and proximal small bowel. There is situs solitus with a left-sided aortic arch. Most likely diagnosis?**

Tracheoesophageal fistula and Vater syndrome.

☐ **If the patient has EA, how did air reach the stomach and proximal small bowel?**

A distal TEF or an H-type fistula allows air to move into the stomach. With an H-type fistula, the tube should not have coiled in the proximal esophagus. Thus, the child probably has EA with a distal TEF.

☐ **A 3-year-old male presents with a palpable right abdominal mass. CT reveals the mass to be located in the region of the right adrenal gland. The mass does not cross the midline but does invade the IVC. There are no calcifications. The child is noted to have aniridia. Differential diagnosis?**

Wilms tumor versus neuroblastoma.

TOXICOLOGY AND ENVIRONMENTAL

☐ **Which metabolic pathway for acetaminophen predominates in childhood?**

Sulfonation.

☐ **At what approximate age does the predominant metabolism of APAP switch from sulfonation to glucuronidation?**

14-year-old.

☐ **What is the volume of distribution for acetaminophen?**

1L/kg.

☐ **What is the toxic dose of acetaminophen in acute ingestion?**

140 mg/kg.

☐ **What percentage of a therapeutic dose of acetaminophen is normally metabolized by the cytochrome P450 system?**

3 to 15%.

☐ **What are the symptoms in the first few hours after an acetaminophen overdose?**

None or mild gastrointestinal symptoms.

☐ **T/F: Acetaminophen overdose may cause a depressed mental status in the first few hours after overdose.**

True. This only occurs in large overdoses.

☐ **What is the oral dose of N-acetylcysteine in acetaminophen overdose?**

140 mg/kg followed by 70 mg/kg every four hours for 17 doses.

☐ **T/F: The dose of N-acetylcysteine should be adjusted if it is given with activated charcoal.**

False.

☐ **T/F: Charcoal hemoperfusion should be used in severe acetaminophen toxicity.**

False. N-acetylcysteine should be utilized since it is an effective antidote.

☐ **For what toxic ingestions is N-acetylcysteine indicated?**

Acetaminophen, pennyroyal, carbon tetrachloride and chloroform.

☐ **What complication may arise from rapid IV administration of N-acetylcysteine?**

Anaphylactoid reaction.

☐ **With what laboratory test may N-acetylcysteine interfere?**

It may cause a false positive for urine ketones.

☐ **T/F: Intravenous N-acetylcysteine is FDA approved for use in the US.**

True. Acetadote received FDA approval.

☐ **What is the toxic metabolite of acetaminophen produced via the cytochrome oxidase P450 isoenzymes?**

N-acetyl-p-benzoquinoneimine (NAPQI).

☐ **Why is excess NAPQI produced in acetaminophen overdose?**

The normal sulfation and glucuronidation pathways are saturated; so more acetaminophen is metabolized to NAPQI by the cytochrome P450 isoenzymes.

☐ **Why does NAPQI accumulate after acetaminophen overdose?**

Hepatic glutathione stores are depleted.

☐ **T/F: Acetaminophen overdose may cause early metabolic acidosis**

True, in severe overdoses.

☐ **What is the most sensitive indicator of liver injury after acetaminophen overdose?**

Aspartate aminotransferase (AST) level.

☐ **How early are elevations of the AST are seen after acetaminophen overdose?**

After 8 to 12 hours in severe toxicity.

TOXICOLOGY AND ENVIRONMENTAL

☐ **When is the period of maximal liver injury after acetaminophen overdose?**

3 to 4 days post-ingestion – stage III.

☐ **What is phase IV of acetaminophen overdose?**

The recovery phase in those who survive phase III.

☐ **Is a 2-hour acetaminophen useful?**

Since the Rumack-Matthew nomogram for acute acetaminophen ingestions starts at four hours, a two hour level is difficult to interpret. However, a very low level at two hours probably rules out a significant ingestion, except potentially in cases where an extended release product was ingested or gut motility is decreased.

☐ **Are children at more or less risk of acetaminophen toxicity?**

Children seem to be at less risk, except in the case of chronic ingestions in febrile infant

☐ **What groups of patients may be at increased risk of acetaminophen toxicity?**

Febrile infants, patients taking medicines which induce cytochrome P450 isoenzymes, the malnourished, and HIV patients.

☐ **How does acetaminophen metabolism in children and adults differ?**

In adults, significantly more acetaminophen undergoes glucuronidation than sulfation, while in children the opposite is true.

☐ **When should NAC therapy be instituted empirically?**

If an acetaminophen level is not available within 8 hours of the ingestion.

☐ **Does activated charcoal administration decrease the half-life of acetaminophen?**

No.

☐ **How may NAC decrease the amount of NAPQI that is generated?**

By enhancing the sulfation pathway of metabolism.

☐ **What is the primary indication for the use of ethosuximide?**

Petit mal seizures.

☐ **What are the anti-convulsant mechanisms of action of valproic acid?**

Na+ channel inactivation and inhibition of GABA breakdown by GABA transaminase resulting in increased GABA.

PEDIATRIC EMERGENCY MEDICINE BOARD REVIEW

☐ **What fatal CNS effect may occur after valproic acid overdose?**

Cerebral edema.

☐ **What patients are at higher risk of valproic acid induced liver failure?**

Children under two and those on multiple anticonvulsants.

☐ **Narcan has been used to reverse the sedation after overdose of which anticonvulsant?**

Valproic acid.

☐ **What is the clinical implication of QRS duration greater than 100 msec in cyclic antidepressant overdose?**

Approximately 1/3 of patients will seize.

☐ **What is the treatment of dystonic reactions?**

Anticholinergic agents, such as diphenhydramine (1-2 mg/kg IV) or benztropine mesylate (1-2mg IV).

☐ **What is the treatment of dystonic reactions if anticholinergic agents are not effective?**

If anticholinergic agents are ineffective, reconsider the diagnosis and treat with benzodiazepines.

☐ **What is the pathophysiology of tardive dyskinesia?**

Chronic blockade of dopamine receptors causes an increase in release of dopamine, particularly in the nigrostriatal neurons.

☐ **Why do tricyclic antidepressants cause seizures in overdose?**

The tricyclic antidepressants bind to the picrotoxin site on the GABA channel and inhibit chloride flow.

☐ **T/F: Patients who ingest large amounts of tricyclic antidepressants and arrive in the emergency department with minimal symptoms usually do not worsen.**

False. In one study, 50% of all patients with mild toxicity rapidly worsened within the first hour.

☐ **Approximately what percentage of patients with tricyclic antidepressant toxicity and a QRS > 100 msec will develop seizures?**

30%.

☐ **Approximately what percentage of patients with tricyclic antidepressant toxicity and a QRS > 160 msec will develop dysrhythmias?**

50%.

TOXICOLOGY AND ENVIRONMENTAL

☐ **What morphologic change in the AVR lead of an ECG is predictive of tricyclic antidepressant toxicity?**

Rightward axis shift in the terminal 40 msec.

☐ **What is the first-line treatment of wide complex tachycardia from tricyclic antidepressant toxicity?**

Sodium bicarbonate.

☐ **What is the second-line treatment of wide complex tachycardia from tricyclic antidepressant toxicity?**

Hypertonic saline.

☐ **What antidotes are contraindicated in tricyclic antidepressant toxicity?**

Physostigmine and flumazenil.

☐ **What is the first-line therapy for hypotension from tricyclic antidepressant toxicity?**

After adequate fluid resuscitation, an alpha agonist such as norepinephrine is most appropriate.

☐ **What are the side effects of SSRIs in therapeutic use?**

Lightheadedness, anorexia and blurred vision. The SSRIs may also cause increases in serum glucose, cortisol, ACTH and vasopressin leading to an SIADH-like syndrome. Platelet dysfunction may also occur.

☐ **What are the clinical signs of serotonin syndrome?**

Mental status changes (agitation, sedation, coma), musculoskeletal irritability (tremor, rigidity, myoclonus or hyperreflexia), autonomic instability and fever.

☐ **What is the cause of serotonin syndrome?**

Excessive stimulation of the 5HT 1A receptor. This may occur particularly when two serotonergic medications are used concurrently, but has been reported with single medications.

☐ **What is the time of onset of serotonin syndrome?**

Minutes to hours after ingestion.

☐ **What is the time of onset of the neuroleptic malignant syndrome (NMS)?**

Generally 3 to 9 days after the neuroleptic dose.

PEDIATRIC EMERGENCY MEDICINE BOARD REVIEW

☐ **What is the antidote used to treat acetaminophen toxicity?**

N-acetylcysteine.

☐ **What is the oral dose of N-acetylcysteine?**

First dose is 140 mg/kg, then 70 mg/kg q4 hours for 17 doses.

☐ **What is the antidote used to treat iron toxicity?**

Deferoxamine.

☐ **What is the dose for deferoxamine?**

15 mg/kg/hr.

☐ **What is the antidote currently used to treat cyanide toxicity in the United States?**

Pasadena Cyanide Kit. This kit includes amyl nitrate, sodium nitrate and sodium thiosulfate.

☐ **What is the antidote used to treat organophosphate toxicity?**

Atropine for muscarinic effects and 2-PAM for nicotinic effects.

☐ **What is the antidote that may be used to treat black widow spider bite related toxicity?**

Latrodectus antivenom.

☐ **What is the dose of black widow antivenom?**

1-2 vials.

☐ **Black widow antivenom is derived from what animal?**

Horse serum.

☐ **What is the antidote that may be used to treat ethylene glycol toxicity?**

Ethanol or 4 methyl pyrazole (fomepizole).

☐ **What is the antidote that may be used to treat methanol toxicity?**

Ethanol or 4 methyl pyrazole (fomepizole).

☐ **What is the elimination half life of naloxone?**

Approximately one hour.

TOXICOLOGY AND ENVIRONMENTAL

☐ **What is the antidote that may be used to treat lead toxicity?**

EDTA, BAL, and DMSA.

☐ **What is the antidote that may be used to treat anticholinergic syndrome related toxicity?**

Physostigmine.

☐ **What is the antidote that may be used to treat INH related toxicity?**

Pyridoxine (Vitamin B6).

☐ **What is the drug that may be used to treat methemoglobinemia?**

Methylene blue.

☐ **What is the antidote that may be used to treat carbon monoxide related toxicity?**

High-flow oxygen.

☐ **What is the antidote that may be used to treat opioid related toxicity?**

Naloxone (Narcan).

☐ **What is the antidote that may be used to treat beta-blocker related toxicity?**

Glucagon.

☐ **What is the action of flumazenil?**

Highly selective competitive inhibitor of CNS benzodiazepine receptors.

☐ **Which route for the administration of N-acetylcysteine is more efficacious; IV or PO?**

They are equally efficacious.

☐ **What are the indications for whole bowel irrigation?**

Sustained release products or slowly dissolving tablets (paint chips, concretions), foreign bodies (crack vials, etc), body packers, and drugs not adsorbed by charcoal (lithium, iron, metals).

☐ **If pralidoxime chloride is administered intravenously too quickly, what complication may occur?**

Hypertension.

☐ **What adverse reaction to NAC is found following IV use but not oral use?**

Anaphylactoid reactions.

PEDIATRIC EMERGENCY MEDICINE BOARD REVIEW

☐ **What carbamate-like drug can be used as a "protectant" against organophosphate or nerve agent toxicity?**

Pyridostigmine bromide.

☐ **For which substances would urinary alkalinization be indicated?**

Aspirin, Chlorpropamide, Phenobarbital.

☐ **What is the mechanism of warfarin's anticoagulant action?**

Warfarin inhibits the reduction of vitamin K and decreases the concentration of the active form of vitamin K necessary for the vitamin-K dependent clotting Factors (II, VII, IX, X) as Vitamine K serves as a cofactor that catalyzes carboxylation of glutamic acid residues on these Factor's proteins.

☐ **What will reverse the action of warfarin?**

Vitamin K and/or administration of fresh frozen plasma (FFP).

☐ **How does vitamin K reverse warfarin-induced coagulopathy?**

Vitamin K provides functional coagulation factors and bypasses the metabolic block that results in the synthesis of defective coagulation factors.

☐ **How long does it take vitamin K to reverse the effects of warfarin?**

Effects begin at 6-8 hours, but do not peak for 1-2 days.

☐ **What can you do for immediate restoration of hemostasis in the case of warfarin overdose?**

Transfusion of fresh frozen plasma.

☐ **What is the major toxicity of warfarin?**

Bleeding.

☐ **What are the effects of angiotensin converting enzyme inhibitor (ACEI) overdose?**

Symptoms after ACEI overdose are unusual and are limited to hypotension.

☐ **What may reverse the hypotension in ACEI overdose?**

Naloxone has been reported to reverse hypotension. ACEIs may inhibit the metabolism of enkephalins leading to opiate-induced hypotension. Hypotension should, of course, be treated with fluids.

TOXICOLOGY AND ENVIRONMENTAL

☐ **What is the difference between a bacteriostatic and bactericidal antibiotic?**

A bacteriostatic drug inhibits bacterial growth while a bactericidal agent will kill the bacteria.

☐ **What is the mechanism of action of the sulfonamides?**

Inhibition of folic acid synthesis mainly by competitive inhibition of the enzyme dihydropteroate synthetase.

☐ **What is the natural metabolite that the sulfonamides compete with?**

PABA (p-aminobenzoic acid).

☐ **Which is the most common adverse effect associated with the use of sulfonamides?**

Allergic reactions especially skin rashes.

☐ **What are the advantages of combining sulfamethoxazole with trimethoprim?**

Bactericidal effect; decreased resistance; less toxicity and an extended spectrum of activity.

☐ **What is the major adverse effect associated with penicillin use?**

Hypersensitivity reactions.

☐ **What toxicities are associated with ampicillin?**

The incidence of rash is high in some patients. Pseudomembranous colitis may raely occur.

☐ **What is the purpose of combining clavulanic acid with penicillin?**

This combination provides a synergistic effect. Clavulanic acid is a beta lactamase inhibitor and thus extends the action of the penicillin.

☐ **What additional toxicity may be associated with the combinations of beta lactamase inhibitors with beta lactam antibiotics?**

GI effects such as diarrhea.

☐ **What is the advantage of amoxicillin over ampicillin?**

Amoxicillin has better oral bioavailability.

☐ **Are amoxicillin and ampicillin inactivated by penicillinase?**

Yes.

☐ **Which group of cephalosporins has the best penetration into the CNS?**

Third generation.

☐ **What is the major adverse effect associated with cephalosporin use?**

Rashes and other hypersensitivity reactions.

☐ **What ribosomal subunit do the macrolide antibiotics bind to?**

50S.

☐ **What other antibiotics bind to this subunit?**

Clindamycin and chloramphenicol.

☐ **What are the advantages of the newer macrolide antibiotics over erythromycin?**

Less GI irritation and better absorption.

☐ **What is the most serious toxicity associated with the use of erythromycin?**

Cholestatic hepatitis.

☐ **Which erythromycin preparation is cholestatic hepatitis most commonly associated with?**

Erythromycin estolate (Ilosone).

☐ **What is the major type of antibacterial activity associated with clindamycin?**

Anaerobic organisms especially Bacteroides fragilis.

☐ **What is the most serious toxicity associated with the use of clindamycin?**

Pseudomembranous colitis (AAPC).

☐ **How is AAPC (Antibiotic-Associated Pseudomembranous Colitis) caused?**

By a toxin secreted by clindamycin resistant strains of C. difficile.

☐ **What drugs are used to treat AAPC?**

Oral vancomycin and oral/IV metronidazole.

TOXICOLOGY AND ENVIRONMENTAL

☐ **What is the common toxicity associated with the use of vancomycin?**

Red neck or red man syndrome.

☐ **What are the two most serious toxicities associated with the use of vancomycin?**

Ototoxicity and nephrotoxicity.

☐ **Describe the mechanism of action of metronidazole?**

The anaerobic environment produces certain chemically active reduction products, which can disrupt DNA and inhibit nucleic acid synthesis.

☐ **What are the common adverse effects of metronidazole?**

GI upset and some neurological toxicity (CNS effects and peripheral neuropathy).

☐ **What is the gray baby syndrome?**

Infants with their limited ability to metabolize chloramphenicol experience a syndrome, which includes an ashen gray cyanosis, which may be fatal.

☐ **Why are the tetracyclines contraindicated in young children.**

They can cause discoloration of the teeth and can inhibit bone growth.

☐ **What is the mechanism of the tetracyclines effect in bones and teeth?**

Tetracyclines bind to calcium and then become deposited in the bones and teeth. Discoloration of the teeth is sometimes due to oxidation of the tetracycline-calcium complex.

☐ **What are the clinical uses of ketoconazole?**

A few systemic fungal infections and dermatophyte infections that are resistant to griseofulvin.

☐ **What adverse reactions are associated with the use of ketoconazole?**

GI intolerance, endocrine effects (decreased libido, oligospermia etc.) and hepatic toxicity.

☐ **Why does pyridoxine prevent the neurotoxicity associated with the use of INH?**

INH causes pyridoxine deficiency by forming hydrazones that inhibit the conversion of pyridoxine to the active form pyridoxal phosphate.

☐ **What is the mechanism of action of rifampin?**

It inhibits mycobacterial DNA-dependent RNA polymerase.

☐ **How does the body excrete rifampin?**

It is rapidly eliminated in the bile, followed by an enterohepatic recirculation. Eventually about 60% is excreted via the intestine.

☐ **How is rifampin metabolized?**

It is progressively diacetylated in the liver with the metabolite retaining full activity. It induces its own metabolism.

☐ **What adverse effects are associated with the use of rifampin?**

Gastrointestinal disturbances and hepatotoxicity.

☐ **What additional toxicity is sometimes seen when rifampin is used in high dose intermittent therapy?**

It produces several effects thought to have an immunological basis, including thrombocytopenia, hemolytic anemia, acute renal failure and the respiratory flu syndrome.

☐ **What type of drug-drug interactions are associated with rifampin use?**

Rifampin is a potent inducer of liver microsomal enzymes and may affect the metabolism of a variety of other drugs.

☐ **Structurally what is acyclovir?**

An analog of guanosine or deoxyguanosine with an acyclic side chain.

☐ **What is the mechanism of action of acyclovir?**

It is phosphorylated to mono, di- and tri- phosphates. As the nucleotide it is an inhibitor of viral DNA polymerase and is incorporated into the growing DNA chain where it can act as a chain terminator due to the lack of the 3'- hydroxyl group.

☐ **What is the most common adverse effect associated with the use of acyclovir?**

Renal toxicity.

☐ **What causes the renal toxicity associated with acyclovir?**

High urine levels of the drug cause crystalline nephropathy.

☐ **What other adverse effects are associated with the use of acyclovir?**

Neurotoxicity is uncommon and GI upset, headache and rash occur when it is given orally. There may also be lacrimation, itching, burning and pain when it is used in the eye.

TOXICOLOGY AND ENVIRONMENTAL

☐ **What is valacyclovir?**

A prodrug of acyclovir that is rapidly converted to acyclovir after absorption. It has a mechanism of action identical to that of acyclovir.

☐ **What adverse effects are associated with the use of valacyclovir?**

It is generally well tolerated, but a thrombotic thrombocytopenic purpura/hemolytic anemia syndrome has been reported in some severely immunocompromised patients treated with high doses for prolonged periods. Hallucinations and confusion also may occur. Other adverse effects are similar to acyclovir.

☐ **What are the toxic effects of quinine and quinidine?**

Cinchonism (tinnitus, headache, nausea, hearing and visual disturbances); hyperinsulinemia and severe hypoglycemia; severe hypotension when given too rapidly by intravenous injection.; overdoses produce more severe symptoms including GI and cardiovascular. disturbances. There may also be hemolysis in glucose-6-phosphate dehydrogenase deficient patients.

☐ **What odor may emanate from an arsenic toxic patient?**

Garlic.

☐ **What is generally thought to be the drug of choice in a patient with a beta-blocker overdose who is bradycardic and hypotensive without response to intravenous fluids and atropine?**

Glucagon.

☐ **Stimulating what other receptor can increase decreased intracellular cAMP from beta-receptor blockade?**

Glucagon.

☐ **What symptom is commonly seen when glucagon is administered rapidly intravenously?**

Vomiting.

☐ **In which of the beta-blocker overdoses should hemodialysis be considered after other measures fail?**

Acebutolol, nadolol, atenolol, sotalol.

☐ **How long should patients with sustained release beta-blocker overdose be observed?**

24 hours.

☐ **What two abnormalities on the chemistry panel may be seen with beta-blocker overdose?**

Hyperkalemia and hypoglycemia.

☐ **Which group of beta-blocker overdosed patients has a higher incidence of hypoglycemia?**

Children.

☐ **What toxin related disease in infants is related to the ingestion of raw honey?**

Infantile botulism.

☐ **What is the sine quo non of botulism poisoning presentation?**

Bulbar palsy.

☐ **Name four sources of botulism.**

Botulism occurs with the ingestion of a preformed toxin in improperly preserved, canned, smoked, or fermented foods, improperly stored fried onions, baked potatoes, potato yogurt sweetened with aspartame, and turkey stuffing.

☐ **Name the bacteria that produces the botulinum neurotoxin.**

Clostridium botulinum.

☐ **How does the Clostridium botulinum cause toxicity?**

The neurotoxin causes motor weakness by binding to the cholinergic nerve terminals of the axon and inhibiting acetylcholine release.

☐ **How does death normally occur from botulism?**

Respiratory muscle failure.

☐ **What is the order of progressive muscle involvement in a patient with botulism?**

Muscles served by cranial nerves, followed by extremity muscles and finally respiratory muscles.

☐ **At what age is botulism most commonly observed in infants?**

1-13 months.

☐ **In what kind of patient should botulism be considered?**

Any patient with a history of gastrointestinal symptoms, dry mouth, difficulty swallowing and visual symptoms has botulism until proven otherwise.

☐ **What samples should be collected from a child suspected to have botulism?**

Serum, stool, gastric contents and suspect foods should be tested for botulinum toxin and Clostridium botulinum.

TOXICOLOGY AND ENVIRONMENTAL

☐ **What type of monitoring does a child suspected of having botulism need?**

Respiratory status must be followed using measurements of vital capacity, peak expiratory flow rate, negative inspiratory force, pulse oximetry, and gag reflex.

☐ **When should a child with botulism be intubated?**

Some sources suggest that intubation should be considered in a child suspected of having botulism when the vital capacity falls to less than 30% of predicted or for any signs of respiratory distress with impending failure.

☐ **What are the common manifestations of infant botulism?**

Constipation, feeding difficulty, weak cry, decreased muscle tone, ophthalmoplegia and loss of facial grimacing can occur.

☐ **How is infant botulism different from that observed in adults?**

In the adult form, the preformed toxin is ingested. The toxin is produced in vivo in cases of infant botulism.

☐ **Why is the bacteria able to produce toxin in infants?**

The infant gut lacks the intestinal flora, gastric acids and bile acids that inhibit bacterial growth in adults.

☐ **What is the most common food associated with infant botulism?**

Honey.

☐ **What is the least common cause of botulism?**

Wound botulism.

☐ **How soon after ingestion of contaminated honey do infants show symptoms of botulism?**

18 to 36 hours after toxin enters the infant's body.

☐ **What strains of Clostridium botulinum are most frequently implicated in human disease?**

Strains A, B, and and E.

☐ **What conditions must exist for botulinum spores to germinate?**

They require an anaerobic environment with a pH greater than 4.6.

☐ **How can botulinum spores be destroyed?**

Pressure cooking for 30 minutes at 120 C will destroy the spores.

☐ **Is the toxin more difficult or less difficult to destroy than the spores?**

The toxin is much more heat labile than the spores. Boiling at 100 C for minute will destroy the toxin.

☐ **What can be added to foods to inhibit Clostridium growth?**

Nitrites.

☐ **How long after ingestion of botulinum do symptoms occur?**

The incubation period is usually 12-36 hours, but can take as long as 8 days.

☐ **Does breastfeeding help protect infants from botulism?**

No, breast fed infants are more likely to develop botulism.

☐ **What subset of patients are at highest risk for developing wound botulism?**

Drug abusers who inject subcutaneously with infected needles, forming abscesses in which the bacterium can produce toxin.

☐ **What botulin antitoxin effectively neutralizes all seven known botulinum nerve toxin serotypes (A-G)?**

The Heptavalent Botulism Antitoxin (HBAT) was approved in 2010 by the CDC.

☐ **How do the antitoxins work?**

They bind circulating free toxin and prevent the progression of illness.

☐ **When is the antitoxin most effective?**

It should be given within 24 hours of the onset of symptoms.

☐ **What test should be performed prior to administering the antitoxin?**

A skin test to rule out hypersensitivity should be performed. Serum sickness is a potential complication because the antitoxin is derived from horse serum.

☐ **Is botulism a reportable disease?**

Yes, botulism should be reported to local or state health departments or the Centers for Disease Control.

☐ **What is the most common form of botulism?**

Infant botulism accounts for two thirds of reported cases.

TOXICOLOGY AND ENVIRONMENTAL

☐ **Where is botulism most commonly found in North America?**

Alaska, California, Oregon, and Washington have the highest incidences of botulism.

☐ **Where are the different toxin types most often found?**

Type A and B are ubiquitous in soil, type E is found in mud and sand in oceans. Type A occurs more commonly west of the Mississippi, type B more often east of the Mississippi.

☐ **Should antibiotics be used in the treatment of botulism?**

Antibiotics should be used only to treat secondary infections.

☐ **What is guanidine's effect in patients with botulism?**

It increases the release of acetylcholine at the nerve terminal but has not been shown to be clinically useful.

☐ **What is the dose of antitoxin in suspected botulism cases?**

1-2 vials.

☐ **Should antitoxin be used in cases of infant botulism?**

No.

☐ **Should cathartics be used after botulism ingestion?**

Cathartics should not be used, especially those containing magnesium because they may exacerbate muscle weakness.

☐ **Which adults are susceptible to an adult-form of infant botulism?**

Adults with bowel pathology such as Crohn's disease or achlorhydria that allows the bacteria to produce toxin.

☐ **What kind of organism is Clostridium botulinum?**

It is a gram-positive, anaerobic organism.

☐ **What pupil findings are suggestive of botulism?**

Fixed or dilated pupils in an alert patient.

☐ **How long do the symptoms of botulism persist?**

Hospitalizations last 1-10 days, but symptoms such as fatigue can last 1-2 years.

☐ **What drug may mildly and transiently improve the symptoms of botulism.**

Edrophonium may improve symptoms slightly, although too much less an extent than in a patient with myasthenia gravis.

☐ **Honey should not be given to children younger than what age?**

1 year.

☐ **How are deep tendon reflexes affected in botulism?**

Reflexes are either normal or hyporeflexic.

☐ **What are the CSF findings in botulism?**

Lumbar puncture usually reveals normal CSF, although occasionally slightly elevated protein may be present.

☐ **What are the indications for hyperbaric oxygen therapy in carbon monoxide poisoning if there is easy access to a hyperbaric chamber?**

COHbg level > 25%, syncope, coma, seizures, altered mental status, myocardial ischemia, focal neurologic deficit, ventricular dysrhythmias, pregnancy with a COHb > 15% or persistent neurologic symptoms after oxygen therapy.

☐ **What percentage of patients with carbon monoxide intoxication will develop delayed neurologic sequelae?**

14-43%.

☐ **When is the onset of delayed neurologic sequelae after carbon monoxide poisoning?**

2-40 days after exposure.

☐ **Carbon monoxide has an affinity for hemoglobin that is greater than oxygen's. How much larger?**

Two hundred and fifty times.

☐ **How many deaths are caused by carbon monoxide poisoning per year?**

CO causes approximately 3000-5000 deaths per year in the US.

☐ **What are the most common symptoms of carbon monoxide toxicity?**

Headache (91%), nausea and vomiting (47%), dizziness and weakness (53%) and dyspnea (40%).

TOXICOLOGY AND ENVIRONMENTAL

☐ **In which time frame do the delayed neurologic sequelae of carbon monoxide occur: first hour after exposure, one to six hours, six to twenty-four hours or two to forty days?**

The delayed neurologic sequelae after carbon monoxide exposure typically develop between two and forty days after the exposure.

☐ **What odor is consistent with ambient levels of carbon monoxide?**

None, it is an odorless gas.

☐ **T/F: Carbon monoxide casualties have resulted from the generation of carbon monoxide as the result of blasting operations?**

True.

☐ **What is the level of CO that is considered immediately dangerous to life and health (IDLH)?**

1200 ppm.

☐ **What is a "normal" serum level of carboxyhemoglobin (COHgb) in a child not exposed to exogenous carbon monoxide?**

Generally below 3%.

☐ **What is the source of the baseline carboxyhemoglobin level in non-exposed patients?**

Carbon monoxide is a byproduct of heme degradation.

☐ **Name four sources of carbon monoxide.**

Incomplete combustion of fossil fuels (propane heaters, boat engines, tractor engines, car engines, generators, gas grills etc), fire/smoke inhalation, cigarette smoke, and methylene chloride.

☐ **How does methylene chloride produce carbon monoxide?**

Methylene chloride is a paint stripper that is easily absorbed from the skin or inhaled. It is metabolized to carbon monoxide in the liver.

☐ **How does carbon monoxide affect hemoglobin oxygen binding?**

CO has an affinity for hemoglobin that is 250 times greater than oxygen. This leads to decreased arterial oxygen content despite a normal pO_2.

☐ **How does carbon monoxide affect the oxygen-hemoglobin dissociation curve?**

Carbon monoxide causes a conformational change in hemoglobin that allows the three remaining oxygen molecules to be held more tightly. This leads to a left shift in the oxygen-hemoglobin dissociation curve.

☐ **What is carbon monoxide's effect on mitochondrial cytochrome oxidase?**

CO binds to cytochromes A3 and P450 and inhibits cellular respiration and ATP production.

☐ **What is carbon monoxide's effect on myoglobin?**

CO binds to myoglobin leading to myocardial impairment. This binding is increased in hypoxic conditions.

☐ **What is carbon monoxide's effect on platelets?**

CO causes displacement of NO from platelets leading to smooth muscle relaxation and vasodilation.

☐ **What clinical signs/symptoms are expected with a COHgb level of 10-20%?**

Headache, nausea, vomiting and dizziness.

☐ **What are the ophthalmologic signs of severe carbon monoxide poisoning?**

Flame hemorrhages, papilledema and bright red retinal veins (equilibration of the color of arteries and veins).

☐ **T/F: Fetal COHgb levels are lower than maternal levels.**

False. Fetal COHgb levels are 10-15% higher than maternal levels.

☐ **T/F: Fetal COHgb levels do not equilibrate with maternal levels for up to 24 hours.**

True. Fetal levels lag behind maternal levels and equilibrate after 14-24 hours of exposure.

☐ **What percentage of children born to women with CO toxicity will have neurologic sequelae?**

Up to 60%.

☐ **What is the complication rate of hyperbaric oxygen therapy?**

Approximately 5%.

☐ **What is the severe complication rate of hyperbaric oxygen therapy?**

Approximately 3%. 1% develop dysrhythmias, 1% develop tension pneumothorax, 1% arrest, 2% develop hypotension and 3% have a seizure. Many of these complications may be related to underlying diseases.

☐ **Can just a small increase in calcium channel blocker dose be potentially lethal?**

Yes.

TOXICOLOGY AND ENVIRONMENTAL

☐ **What rhythm abnormality results when a calcium antagonist binds the calcium channel receptors in the sinoatrial node?**

Sinus bradycardia, pause, or arrest.

☐ **Is hyperglycemia a common finding in calcium channel blocker overdose?**

Yes. Probably because there is calcium mediated release of insulin from the beta cells of the pancreas that may be blocked by high concentrations of calcium channel blockers.

☐ **Can nausea/vomiting be seen in calcium channel blocker overdose?**

Yes.

☐ **Are CNS changes a prominent feature of calcium channel blocker poisoning independent of blood pressure and cardiac output changes?**

No.

☐ **What is the initial treatment (other than intravenous saline and atropine) for patients that have taken a calcium channel blocker overdose and are bradycardic and hypotensive?**

Intravenous calcium.

☐ **What is the difference between calcium chloride and calcium gluconate?**

Calcium chloride is 3 times as concentrated as calcium gluconate.

☐ **What is the decontamination method of choice for an overdose of a sustained release calcium channel blocker presenting greater than 1 hour post-ingestion.**

Polyethylene glycol (Go-lytely).

☐ **If there is no effect in treating a patient with calcium channel blocker toxicity that is bradycardic and hypotensive with calcium, atropine, and catecholamines, which drug should be considered?**

Glucagon.

☐ **Can catecholamines be used to treat bradycardia and hypotension?**

Yes.

☐ **In biochemical terms, why may glucagon work to reverse the effects of calcium channel blockade?**

It increases calcium entry into the heart cells via increased cAMP and protein kinase activity.

PEDIATRIC EMERGENCY MEDICINE BOARD REVIEW

☐ **What treatment that increases "substrate" delivery to the myocardium is very promising.**

Insulin and glucose (hyperinsulinemic euglycemia treatment).

☐ **If insulin and glucose are administered, what electrolyte must be closely observed?**

Potassium, as these drugs force potassium intracellularly.

☐ **Is hemodialysis or hemoperfusion likely to enhance the elimination of calcium channel blockers?**

No.

☐ **Is it true that sustained release calcium channel blockers (e.g. verapamil SR, diltiazem CD, nifedipine XL) are "long acting" because the medication remains in the gut and slowly releases the medication**

Yes.

☐ **What implications does this have in these overdoses regarding GI decontamination?**

Further absorption can theoretically be decreased by GI decontamination even 1-2 hours after ingestion with repeated dose activated charcoal or whole bowel irrigation.

☐ **Are symptoms from sustained release calcium channel blocker overdose, e.g. verapamil SR, generally more delayed than regular verapamil?**

Yes.

☐ **What is the initial dose of glucagon in calcium channel antagonist toxicity?**

2-5 mg over 1-3 minutes in adults and 50 mcg/kg in children.

☐ **Can ingestion of small amounts of household bleach (2 teaspoons) cause severe gastrointestinal irritation?**

No.

☐ **What sort of damage is seen on endoscopy with bleach ingestion?**

Endoscopy is indicated if there are severe symptoms such as pain, vomiting, hematemesis. Usually there is mucosal irritation, which appears as a white film on the esophageal and gastric mucosa. There is rarely an ulcer.

☐ **How about the patient that drinks "Drano"? Is this a serious ingestion?**

Yes.

TOXICOLOGY AND ENVIRONMENTAL

- **What type of necrosis does "Drano" cause?**

    Liquefactive necrosis.

- **What is the significance of this type of necrosis?**

    Liquefactive necrosis allows the material, in this case a strong base, to penetrate deeply into the tissue, such as the esophageal wall, resulting in a thick layer of injury. Thus there is a larger chance of perforation, and if perforation does not occur immediately, there is a high chance of stricture formation.

- **Should a child that drank "Drano" undergo gastric lavage, or induction of vomiting?**

    No. There is more danger when the strong base comes back up the esophagus.

- **Should the child that drinks "Drano" be given charcoal?**

    No. Charcoal may also cause vomiting, and there is nothing to be gained by preventing the absorption of lye as its toxicity from local contact. It will also obscure the endoscopist's view.

- **What should be done if a child drinks "Drano"?**

    You can try to dilute the ingested material in the stomach by asking the patient to drink water or milk. If the ingestion was immediate, there is a report of immediately drinking a mild acid such as Coca-Cola (Phosphoric acid), which resulted in a better outcome than drinking water.

- **What type of chemical (acid or base) is found in most toilet bowl cleaners?**

    Acid.

- **What type of "burn" does this type of material cause if exposed in high concentration?**

    Coagulative necrosis.

- **What is the significance of this type of necrosis on long term sequelae?**

    Because the coagulation of tissue proteins occurs by contact, the acid is prevented from reaching the deeper layers of the tissue, so injury, which can appear horrible, is limited to the more superficial layers, and therefore there are fewer long term effects than one would see with burns by a strong base.

- **Which is the more dangerous inhalation household bleach or swimming pool disinfectants?**

    Swimming pool disinfectants, as these may be up a 20% concentration while household bleach is usually a 3-5% concentration.

- **What are soaps?**

    Salts of fatty acid.

PEDIATRIC EMERGENCY MEDICINE BOARD REVIEW

☐ **What symptoms do they produce when acutely ingested?**

Nausea, and vomiting.

☐ **What symptom may develop the next day after soap ingestion?**

Diarrhea.

☐ **What is the difference between soap and a detergent?**

A detergent contains surfactant.

☐ **What does the class of material in detergents allow?**

Reduction of surface tension, allowing easier wetting of surfaces.

☐ **What is the main toxicity of detergents?**

Mucous membrane irritation, skin irritation, possibly superficial mucosal erosion.

☐ **Is the dishwasher detergent sludge remaining in the dispenser toxic?**

Yes, because it is very concentrated.

☐ **Is inhaling detergent toxic?**

Yes. Inhaling powdered detergent can result in respiratory problems the most severe of which is upper airway obstruction (drooling, stridor) which may require intubation.

☐ **What is additional toxicity of "enzyme-containing" detergents?**

As these "enzymes" are proteins, they can cause skin sensitization and allergic reactions, which is some patients may even result in bronchospasm.

☐ **Can there be tetany and hypocalcemia resulting from ingestion of large amounts of phosphate containing detergents?**

Yes, theoretically.

☐ **Has there ever been a death from detergent ingestion?**

Yes. Once exceedingly rare, they have become much more common with the entry of detergent "pods" into the marketplace.

TOXICOLOGY AND ENVIRONMENTAL

☐ **"Pine sol" disinfectant contains an anionic detergent, isopropyl alcohol, and pine oil. Should vomiting or gastric lavage be performed?**

No. Anionic detergents are not seriously toxic. Isopropyl alcohol in the amount that is most likely going to be ingested is not seriously toxic. The pine oil can cause chemical pneumonitis as it is low viscosity and all attempts should be made to prevent aspiration.

☐ **Phenol was originally used in many households (and hospitals) as a potent germicidal agent, but now found mostly in topical skin preparations such as _____.**

Campho-Phenique, and commercial skin pealing agents.

☐ **Creosote, creosol, phenylphenol, hydroquinone, eugenol are derivative of what chemical?**

Phenol.

☐ **Is "Lysol," a common household disinfectant a phenolic compound?**

Yes. It is bisphenol. It denatures proteins and in the case of bacterial disrupts their cell walls, killing them. In humans it produces a coagulation necrosis and thus corrosive injury.

☐ **What skin exposure characteristically caused painless white "burns".**

Phenol exposure. It then turns red and brown.

☐ **By what two mechanisms do mercuric chloride button batteries lodged in the esophagus produce injury? What is a potential catastrophic complication of esophageal placement?**

1. Leakage of the corrosive metallic salt.
2. Local discharge of electrical current at the site of impaction.

The worst complication is perforation, which on plain x-ray may appear as free air in the mediastinum.

☐ **Is it necessary to remove the button battery if it is seen in the esophagus? How about if it is under the diaphragm? How is removal accomplished?**

Yes if it is in the esophagus, no if it is past the lower esophageal sphincter. Removal is accomplished by endoscopy.

☐ **Naphthalene used to be the active ingredient in mothball. What is it now?**

Paradichlorobenzene.

☐ **What was the toxicity of naphthalene?**

Nausea, vomiting with ingestion; mucous membrane irritation with inhalation. CNS effects ranging from agitation, lethargy, and seizures. In patients with G6PD deficiency, severe hemolysis can result.

PEDIATRIC EMERGENCY MEDICINE BOARD REVIEW

☐ **Is paradichlorobenzene toxic?**

Yes, but not nearly as toxic as naphthalene. Has produced liver necrosis and tremors in animals.

☐ **Is there any way to distinguish naphthalene and paradichlorobenzene?**

Physically there are very difficult to distinguish. Paradichlorobenzene is radiopaque which naphthalene is not.

☐ **Should you administer milk, or other fatty substances, to a patient who acutely ingested naphthalene or paradichlorobenzene in an effort to dilute it.**

No. These substances may enhance absorption.

☐ **What are the active ingredients in deodorants? Are they toxic?**

Aluminum and Zinc, low toxicity.

☐ **What is the active ingredient in fingernail polish or fingernail polish remover?**

These are generally hydrocarbon solvents such as acetone, acetates, toluene, and aromatic hydrocarbons.

☐ **What toxic chemical may artificial-fingernail remover contains?**

Acetonitrile (vinyl cyanide). This breaks down in vivo to form cyanide. Some artificial-fingernail removers contain nitromethane.

☐ **What is the active ingredient in "nail primer" or artificial nail "extender"?**

Methacrylic acid (2-methypropanenoic acid). Has a bad odor and is a caustic. It is also used in bone cements. Inhalation can result in respiratory distress. It is non-toxic once polymerized.

☐ **How long should a patient with a skeletal muscle relaxant overdose be observed?**

At least 6 hours because of the possibility of delayed absorption.

☐ **Is Flumazenil useful in "reviving" patients with skeletal muscle relaxant overdose?**

No.

☐ **Is activated charcoal useful in muscle relaxant overdose?**

Yes.

☐ **Is hemodialysis or hemoperfusion useful?**

No, because there is extensive tissue distribution.

TOXICOLOGY AND ENVIRONMENTAL

☐ **What is the onset of action of Dig-specific antibodies?**

Onset averages 19 minutes after infusion, however peak effect is approximately 4 hours.

☐ **How is the dose of digoxin-specific antibodies calculated?**

# of vials of Digoxin-Fab = [serum digoxin (mg/dL) x Body weight (kg)] / 100.

☐ **What is the volume of distribution for digoxin?**

5-6 L/kg.

☐ **What is the effect of toxic levels of digitalis?**

The sodium/potassium ATPase pump is inhibited, allowing no potassium to be transported into the cells.

☐ **What is the most common rhythm disturbance seen in digitalis toxicity?**

Premature ventricular contractions (PVCs).

☐ **Toxic levels of digitalis acting at what anatomic site in the heart cause premature ventricular contractions (PVC s)?**

The Purkinje fibers.

☐ **How do toxic levels of digitalis affect the Purkinje fibers of the heart?**

Toxic levels of digitalis decrease resting potential and conduction velocity, increase muscle fiber sensitivity to electrical stimuli, and increase automaticity.

☐ **Name four or more plants in which cardiac glycosides can be found?**

Oleander, foxglove, lily of the valley, red squill, and rhododendron.

☐ **What constitutes a toxic dose of digitalis?**

Acute ingestions of 1mg of digitalis in children and 3mg of digitalis can produce elevated serum levels.

☐ **Are children or adults more resistant to digitalis overdose?**

Children appear to have less cardiotoxicity from digitalis overdose than adults do.

☐ **How does the volume of distribution of digoxin compare to digitoxin?**

The volume of distribution of digoxin is large (5-10 L/kg) compared to digitoxin (0.5 L/kg).

PEDIATRIC EMERGENCY MEDICINE BOARD REVIEW

☐ **What is the elimination half-life of digoxin?**

30-50 hours.

☐ **Why is the elimination half-life of digitoxin 5-8 days compared to 30-50 hours for digoxin?**

Digitoxin undergoes more extensive enterohepatic circulation than digoxin.

☐ **How much foxglove or oleander must be ingested to reach toxic levels of digitalis?**

Even a few leaves of either foxglove or oleander are enough to cause toxicity.

☐ **How does the use of digitalis specific antibodies affect the measurement of serum digoxin levels?**

The use of Fab fragments of digoxin cause a falsely elevated serum digoxin level.

☐ **When do peak levels occur after digoxin ingestion?**

About 1.5 to 2 hours.

☐ **When does digoxin reach steady state after ingestion?**

6 to 8 hours.

☐ **Is it the peak or steady state level that correlates with toxicity in digoxin overdose?**

The steady state level should be used to determine toxicity and dosages of Fab fragments.

☐ **What are the most common symptoms in digitalis intoxication?**

Fatigue, nausea, visual disturbances and anorexia occur in 80% of cases. Palpitations can occur due to PVCs.

☐ **Above what potassium level is the administration of digitalis specific antibodies recommended?**

Greater than 5 mEq/L.

☐ **How are digitalis antibodies produced?**

It is derived from sheep immunized with digitalis and the IgG is split with papain.

☐ **What classic visual disturbance occurs in patients with digitalis toxicity?**

Patients report seeing a yellow-green halo around objects.

TOXICOLOGY AND ENVIRONMENTAL

☐ **Which medications should NOT be used to treat ventricular arrhythmias thought to be due to digitalis toxicity?**

Quinidine, procainamide, and bretylium may worsen the dysrhythmia.

☐ **Ventricular dysrhythmias due to digitalis toxicity may respond to what medications?**

Lidocaine and phenytoin.

☐ **Why are phenytoin and lidocaine indicated for the treatment of dysrhythmias secondary to digitalis intoxication?**

They depress ventricular automaticity without slowing AV nodal conduction.

☐ **Why should cardiac pacing be used with extreme caution in patients with digitalis toxicity?**

Pacing induced dysrhythmias are common, occurring in one study in 36% of patients with 13% morality in one study.

☐ **Should cardioversion be used in life-threatening dysrhythmias caused by digitalis toxicity?**

Cardioversion should only be used when all other methods fail to control the dysrhythmia because malignant ventricular dysrhythmia may occur.

☐ **What are possible side effects of digoxin-specific antibody therapy?**

Hypokalemia, worsening congestive heart failure, and rash have been reported in treated patients.

☐ **Why is the measurement of digoxin levels after digoxin specific antibodies not clinically useful?**

Most laboratories cannot measure free digoxin levels. Therefore, the level obtained after antibody administration will represent free plus bound digoxin.

☐ **How is digitalis eliminated?**

Renal. It is filtered by the glomeruli and secreted by the renal tubules.

☐ **What are the indications for Digoxin-specific antibodies in acute digoxin toxicity?**

Hyperkalemia (>5.0 mg/dL), potentially lethal arrhythmias, progressive refractory bradydysrhythmia or highly elevated digoxin levels (>10 ng/mL at steady state or >15 ng/mL at any time).

☐ **Which subunit of the Na/K ATPase does digoxin bind?**

Alpha.

☐ **What is the mortality rate of patients with acute digoxin toxicity and a serum potassium between 5.0 mEq/L and 5.5 mEq/L?**

50%.

☐ **What is the mortality rate of patients with acute digoxin toxicity and a serum potassium above 6.5 mEq/L?**

Approaching 100%.

☐ **What is the volume of distribution of digoxin in children?**

16 L/kg in children.

☐ **Is hemodialysis effective in eliminating digoxin?**

No.

☐ **What medication should be avoided in the treatment of digoxin-related bradycardia?**

Isoproterenol may increase ventricular ectopy and is therefore contraindicated.

☐ **What enzyme catalyzes the synthesis of acetylcholine?**

Choline acetyltransferase.

☐ **Where is acetylcholine stored?**

In the pre-synaptic nerve terminals.

☐ **Acetylcholine (Ach) is the transmitter at what sites?**

At preganglionic fibers to all ganglia; postganglionic parasympathetic fibers to smooth muscle, heart and glands; postganglionic sympathetic fibers to sweat glands; motor fibers to skeletal muscle and in the CNS.

☐ **How do autonomic drugs influence neurotransmission?**

They can alter synthesis of the transmitter, modify the storage mechanism for neurotransmitters, change transmitter release, modify the inactivation mechanism of the transmitter and activate or block receptors. The latter is the most common mechanism.

☐ **What are the steps in synaptic transmission?**

Depolarization of presynaptic terminals, calcium influx and transmitter release. Combination of the transmitter with the postsynaptic receptor, receptor response, destruction or dissipation of the transmitter and repolarization of the postsynaptic membrane.

TOXICOLOGY AND ENVIRONMENTAL

☐ **How is acetylcholine released from terminal vesicles?**

An action potential induces an influx of calcium and initiates of exocytosis.

☐ **How does botulinum toxin work?**

It inhibits the release of acetylcholine.

☐ **What is meant by the term cholinergic crisis?**

This is an effect caused by excessive doses of AchE inhibitors. Symptoms include miosis, sweating, salivation, lacrimation, a hyperactive bowel, muscle weakness and paralysis.

☐ **What determines the duration of cholinesterase inhibition?**

The type of bond formed between the drug and the enzyme.

☐ **What are the therapeutic uses of physostigmine?**

It is used in the treatment of wide-angle glaucoma and for the reversal of anti-cholinergic toxicity.

☐ **Name some insecticides that are anticholinesterase agents?**

Parathion, Malathion, diazinon, Fenton and chlorpyrifos.

☐ **What is the toxicological significance of parathion?**

It may be involved in accidental poisoning, suicide and homicide.

☐ **What are the symptoms of parathion toxicity?**

There are nicotinic signs (muscle weakness and fatigability) and muscarinic signs (salivation, GI hyperactivity, bronchoconstriction etc.) as well as CNS effects such as confusion, ataxia, slurred speech, loss of reflexes, generalized convulsions and respiratory paralysis.

☐ **How is parathion poisoning as well as poisoning with other irreversible cholinesterases treated?**

It depends on how soon after the patient is seen. The area should be decontaminated then respiration supported and then atropine and pralidoxime can be used as antidotes.

☐ **What is pralidoxime?**

A cholinesterase reactivator.

☐ **What is the rationale for the use of pralidoxime?**

It is very reactive with phosphorylated enzymes. It is causes reactivation of phosphorylated acetylcholinesterase.

☐ **How does pyridostigmine compare to other cholinesterase inhibitors?**

It has a longer duration of action than neostigmine.

☐ **What are the signs of mushroom poisoning due to muscarine-containing mushrooms?**

The usual signs of muscarine excess including nausea, vomiting, diarrhea, vasodilatation, reflex tachycardia, sweating, salivation and possible bronchoconstriction.

☐ **How is mushroom poisoning due to muscarine-containing mushrooms treated?**

It is treated with atropine.

☐ **What are the belladonna alkaloids?**

A group of alkaloids isolated from plants related to deadly nightshade. They have antimuscarinic effects.

☐ **Name the belladonna alkaloids?**

Atropine and scopolamine.

☐ **What are the sites of action of atropine?**

It acts at muscarinic receptors, including those in blood vessels, sweat glands and the CNS.

☐ **What is the mechanism of action of atropine and other belladonna alkaloids?**

They antagonize the action of Ach at muscarinic cholinergic receptors. Only at large doses do they act at nicotinic cholinergic receptors.

☐ **What is the order in which different tissues are effected by increasing doses of atropine?**

Receptor blockade depends on the dose of atropine. Salivary and bronchial secretions and sweating are depressed first. Then mydriasis, cycloplegia and vagal blockade of the heart occur, then tone and motility of the GI tract and urinary bladder are affected and finally gastric secretion is partially inhibited.

☐ **How does atropine effect the CNS?**

At a low doses it causes mild vagal stimulation and at larger doses it causes restlessness, excitement, irritability and hallucinations.

☐ **What are the ocular effects of atropine?**

Mydriasis, cycloplegia and photophobia.

TOXICOLOGY AND ENVIRONMENTAL

☐ **Why does atropine cause tachycardia?**

It occurs as a result of vagal blockade.

☐ **What is the effect of atropine on the respiratory system?**

It decreases bronchial secretions and relaxes bronchial smooth muscle.

☐ **What are the GI effects of atropine?**

It inhibits salivary secretions and reduces the tone and motility of the GI tract. There is also a partial reduction of gastric secretion.

☐ **What are the therapeutic uses of atropine?**

It is used in preoperative medication, in bradycardia, in ophthalmology to produce mydriasis, to reduce cholinergic hyperactivity. It is also used to treat Parkinsonism, as an antispasmodic, and in some OTC preparations for rhinitis and hay fever.

☐ **What are the adverse effects of atropine?**

Dry mouth, blurred vision, photophobia, tachycardia, GI distress and hot and dry skin.

☐ **What groups of patients are very susceptible to atropine toxicity?**

Children, especially infants, are very sensitive to the hyperthermic effects of atropine. Deaths have followed doses as small as 2 mg.

☐ **What is another name for adrenergic agonists?**

Sympathomimetics.

☐ **What types of sympathomimetics are there?**

Direct agents, indirectly acting agents, mixed agents and those that act reflexly by stimulation of alpha or beta adrenergic receptors in the circulation.

☐ **What is meant by the term indirect acting agent?**

A drug that acts as a result of releasing norepinephrine from the nerve ending.

☐ **Name two indirect acting adrenergic agonists?**

Tyramine and amphetamine.

PEDIATRIC EMERGENCY MEDICINE BOARD REVIEW

☐ **What organ systems does epinephrine exert its prominent effects on?**

The cardiovascular system (vasoconstriction of arterioles in the skin; viscera and mucus membranes) and the respiratory system. In addition it has certain metabolic effects such as increased glycogenolysis and release of glucagon and decreased release of insulin which results in hyperglycemia.

☐ **What are the cardiac effects of epinephrine?**

There is an increased heart rate and contractility.

☐ **What are the vascular effects of epinephrine?**

Cutaneous blood flow is reduced, renal blood flow is reduced, skeletal blood flow can be increased or decreased, and coronary blood flow is increased.

☐ **How is blood pressure effected by epinephrine?**

There is increased systolic pressure, decreased diastolic pressure and no change in mean pressure.

☐ **How is blood pressure effected by norepinephrine (Levophed)?**

There is increased systolic, diastolic and mean blood pressure.

☐ **What are the effects of epinephrine on the respiratory tract?**

There is bronchodilation.

☐ **What are the metabolic effects of epinephrine?**

There is increased glycogenolysis and release of glucagon and decreased release of insulin resulting in hyperglycemia.

☐ **What are the adverse effects of epinephrine?**

Cardiac arrhythmias, hypertension, palpitations, dizziness, anxiety, headache, tremor, myocardial infarction due to increased cardiac work and pulmonary edema.

☐ **What are the toxic effects of amphetamine?**

Psychological and physical dependence, psychosis, confusion, insomnia, headache, restlessness, heart palpitations, tachycardia and impotence.

☐ **What effects of amphetamine show tolerance?**

The anorexigenic and mood improvement effects.

TOXICOLOGY AND ENVIRONMENTAL

☐ **Name two uses of amphetamine?**

Attention deficit hyperactivity syndrome and narcolepsy.

☐ **What are the pharmacological effects of phenylpropanolamine?**

It causes vasoconstriction and stimulation of the hypothalamic satiety center.

☐ **What are the therapeutic uses of phenylpropanolamine?**

It is used as a nasal decongestant and an appetite suppressant.

☐ **What is the therapeutic use of terbutaline?**

It is a bronchodilator used to treat asthma.

☐ **What is propranolol?**

A nonselective beta adrenergic blocker.

☐ **Describe the cardiovascular effects of propranolol?**

The heart rate decreases, cardiac output decreases, oxygen consumption decreases, there is a slowly developing reduction in blood pressure and sinus rate is decreased as well as conduction is slowed.

☐ **What are the metabolic effects of propranolol?**

The normal rise in plasma free fatty acids associated with the beta response is inhibited and the hyperglycemic response to epinephrine is reduced.

☐ **Does propranolol effect bronchial smooth muscle?**

It increases the tone of bronchial smooth muscle and airway resistance is increased.

☐ **What are the side effects of propranolol?**

There is fatigue, lethargy, GI upset, wheezing and coldness in the extremities.

☐ **What adverse effects are associated with the use of nonselective beta blockers?**

Bradycardia, bronchoconstriction, they may hide the warning signs of hypoglycemia, fatigue, depression and sexual dysfunction.]

☐ **What is the mechanism of action of metoprolol?**

It is a competitive blocker of beta1 receptors with no agonist properties.

☐ **What are the cardiovascular effects of metoprolol?**

It inhibits the inotropic and chronotropic actions of the beta2-adrenergic agonists and it reduces blood pressure.

☐ **How is phenoxybenzamine used clinically?**

It is used for pheochromocytoma and benign prostatic hypertrophy.

☐ **What are the toxicities associated with the use of phenoxybenzamine?**

Postural hypotension, reflex tachycardia and nasal stuffiness.

☐ **What is methylphenidate?**

An amphetamine variant whose major pharmacologic effects and abuse potential are similar to those of amphetamine. It is a mild CNS stimulant with more prominent effects on mental than motor activities.

☐ **What is the therapeutic use of methylphenidate?**

It is effective in treating attention deficit hyperactivity disorder and narcolepsy.

☐ **What is the toxicity of methylphenidate?**

Large doses produce signs of generalized CNS stimulation that may lead to convulsions. It shares the abuse potential of the amphetamines.

☐ **What is an LD50?**

The lethal dose to 50% of the population. It is analogous to the ED50.

☐ **What is meant by NOAEL and NOEL?**

No observed adverse effect level or no observed effect level.

☐ **What is the appropriate dose of polyethylene glycol for whole bowel irrigation in a child?**

0.5-1.0 L/hour.

☐ **What substance is the most common cause of toxin related death?**

Carbon monoxide.

☐ **What toxin causes more deaths in the US than any other?**

Carbon monoxide – over 5000 deaths per year in the US.

TOXICOLOGY AND ENVIRONMENTAL

☐ **What factors influence the ability to remove a substance from the body by the use of hemodialysis?**

Molecular size, volume of distribution, degree of protein binding.

☐ **What are the most common exposures in children below the age of six?**

According to theNPDS data from 2014, the most common exposures are the following (yearly average in parenthesis):
1. Cosmetics/personal care items (150,530)
2. Cleaning substances (118,207)
3. Analgesic agents (100,399)
4. Plants (80,250)
5. Foreign bodies/Toys/Miscelleous (72,099)
6. Topical agents (62,053)

☐ **What is the most common etiology of toxin induced death in all age groups ?**

Carbon monoxide.

☐ **When is peritoneal dialysis indicated in the treatment of the poisoned patient?**

Virtually never.

☐ **What are the causes of high anion gap metabolic acidosis?**

AMUDPILES:
Acetaminophen (massive), Acids (propionic acid – ibuprofen, valproic acid).
Methanol, metformin.
Uremia.
DKA, AKA.
Paraldehyde, Phenformin.
Iron, INH.
Lactic acidosis.
Ethylene glycol.
Salicylates.
Caffeine, CO, CN.
Theophylline.

☐ **What is osmolality?**

Osmolality is the molal concentration of solutes. This is the number of particles in one kilogram of solution.

☐ **How should the elevated temperature associated with toxic (agitated delirium) be treated?**

Pack the patient in ice as quickly as possible.

PEDIATRIC EMERGENCY MEDICINE BOARD REVIEW

☐ **How is osmolality measured?**

Generally by freezing point depression. Vapor pressure method is associated with more falsely low measurements.

☐ **What is osmolarity?**

Osmolarity is the molar concentration of solutes. This is the number of particles in a liter of solution. This value is generally calculated: $2*Na + BUN/2.8 + Glucose/18 + Etoh/4.6$

☐ **What is the surface area of activated charcoal? Super-activated charcoal?**

Activated charcoal has a surface area of 950 m2/g and super-activated charcoal has a surface area of 2000 m2/g.

☐ **What is an anaphylactoid reaction?**

An anaphylactoid reaction is very similar to anaphylaxis; however, it does not require prior exposure to the reaction product. Cases may occur from a radiopaque contrast media or from medications, such as NSAIDs.

☐ **Which drugs commonly cause bradycardia?**

ß-blockers, cardiac glycosides, pilocarpine, and cholinesterase inhibitors (such as organophosphates) are responsible for bradycardia. Sympathomimetics, such as amphetamines and cocaine, and anticholinergics, such as atropine and cyclic antidepressants, commonly cause tachycardia.

☐ **What drug commonly causes both horizontal and vertical nystagmus?**

Phencyclidine (PCP).

☐ **Name 5 drugs or conditions that cause hypertension and/or tachycardia.**

Sympathomimetics.
Withdrawal.
Anticholinergics.
MAO Inhibitors.
Phencyclidine (PCP).
A mnemonic for this is SWAMP

☐ **Name six common drugs that can cause hyperthermia.**

Salicylates.
Anticholinergics.
Neuroleptics
Dinitrophenols.
Sympathomimetics and PCP.
The mnemonic for this one is SANDS-PCP.

TOXICOLOGY AND ENVIRONMENTAL

☐ **Which drugs or environmental exposures can induce bullous lesion formation?**

Sedative hypnotics, carbon monoxide, snake bite, spider bite, caustic agents, and hydrocarbons.

☐ **What is a mnemonic for remembering the drugs that cause nystagmus?**

MALES TIP:
M = methanol.
A = alcohol.
L = lithium.
E = ethylene glycol.
S = sedative hypnotics and solvents.
T = thiamine depletion and Tegretol (carbamazepine).
I = isopropanol.
P = PCP and phenytoin.

☐ **What is a mnemonic for remembering drugs that are radiopaque?**

BAT CHIPS
B = barium.
A = antihistamines.
T = tricyclic antidepressants.
C = chloral hydrate, calcium, cocaine.
H = heavy metals.
I = iodine.
P = phenothiazine, potassium.
S = slow-release (enteric coated).

☐ **Name some side-effects of alkalization of the urine.**

Hypernatremia and hyperosmolality.

☐ **What is the mechanism of action of syrup of ipecac?**

Induces vomiting by stimulation of the chemotactic trigger zone in the brain and causes local stomach irritation.

☐ **What are the components of Lomotil?**

Atropine and diphenoxylate.

☐ **Which anti-diarrheal contains difenoxin and atropine?**

Motofen.

☐ **What is loperamide?**

A meperidine analogue sold over the counter as Imodium, an anti-diarrheal.

PEDIATRIC EMERGENCY MEDICINE BOARD REVIEW

☐ **Why may the toxic effects of Lomotil be delayed several hours or more?**

Since Lomotil slows gastrointestinal transit time, its absorption may be delayed. Also, the more-potent opioid metabolite difenoxin may accumulate as it has a longer serum half-life.

☐ **What are the indications for deferoxamine therapy in iron poisoning?**

Serum iron greater than 400-500 mcg/dL or serious clinical signs of toxicity (shock, acidosis, severe gastroenteritis).

☐ **What blood tube causes false depression of iron levels?**

Lavender tubes contain EDTA and falsely decrease the iron levels.

☐ **What percent of ferrous sulfate tablets are elemental iron?**

20%.

☐ **Infection with what organism is likely in patients who are iron overloaded?**

Yersinia enterocolitica.

☐ **How much iron will be bound by 100 mg of deferoxamine?**

9.3 mg.

☐ **What is the toxic dose of iron?**

Severe toxicity is rare with ingestions of less than 40 mg/Kg however mild symptoms such as nausea may occur at lower doses.

☐ **How much elemental iron is present in ferrous gluconate containing preparations?**

12%.

☐ **How much elemental iron is present in ferrous fumarate containing preparations?**

33%.

☐ **The red tinted urine that results from treatment of iron overdose with deferoxamine is known as?**

Vin rose.

☐ **What is the acid-base disturbance that occurs with iron toxicity?**

Metabolic acidosis.

TOXICOLOGY AND ENVIRONMENTAL

☐ **What is the etiology of acidosis in iron toxicity?**

Iron is converted from ferrous to ferric (Fe^{2+} to Fe^{3+}) and a hydrogen ion is released. Iron also causes lactic acidosis by inhibiting oxidative phosphorylation and increasing anaerobic metabolism by causing hypotension.

☐ **What is iron's effect on coagulation?**

Iron inhibits thrombin formation and activity.

☐ **What x-ray may be helpful in suspected iron toxicity?**

Abdominal x-ray may show radiopaque pills.

☐ **What pattern of liver injury occurs with iron toxicity?**

Periportal hemorrhagic necrosis.

☐ **Why may oral deferoxamine administration after iron overdose worsen the toxicity?**

Deferoxamine combines with iron to form ferrioxamine, which is better absorbed than elemental iron, thus increasing total iron absorption.

☐ **What is the first stage of iron toxicity?**

Iron overdose causes acute GI irritation resulting in abdominal pain, nausea, vomiting, and diarrhea. Iron is corrosive to the GI mucosa and may cause significant gastrointestinal ulceration, hemorrhage and necrosis.

☐ **What is the second stage of iron toxicity?**

The patient may experience a "latent" period after the initial gastrointestinal effect resolve.

☐ **What is the third stage of iron toxicity?**

Patients with severe toxicity will develop shock and worsening metabolic acidosis.

☐ **What is the fourth stage of iron toxicity?**

Hepatic failure.

☐ **What is the fifth stage of iron toxicity?**

The corrosive effects of the iron on the gastrointestinal system may result in strictures.

☐ **T/F: Measurement of total iron binding capacity (TIBC) is useful in predicting iron toxicity after overdose.**

False.

☐ **What is the major side effect of IV deferoxamine administration?**

Hypotension.

☐ **What are the indications for deferoxamine therapy after iron overdose?**

Serum iron level above 500 µg/dL or significant signs of toxicity (shock, acidosis).

☐ **How long should an asymptomatic patient be observed after suspected iron overdose?**

6 hours.

☐ **How long after acute overdose does the peak blood level occur?**

Approximately 4.5 hours.

☐ **What change in the urine may be seen after deferoxamine treatment?**

A "vin rose" color caused by elimination of deferoxamine.

☐ **At what rate is deferoxamine infused?**

15 mg/hg/hr.

☐ **What toxicity may occur after greater than 24 hours of deferoxamine therapy?**

ARDS.

☐ **T/F: Any patient with significant iron toxicity will manifest a WBC count greater than 15,000 and/or blood glucose greater than 150 mg/dL.**

False. These parameters have been associated with elevated iron levels but are not accurate predictors.

☐ **T/F: Iron is not significantly adsorbed by activated charcoal.**

True.

☐ **What dose of ingested iron will generally cause severe toxicity?**

60 mg/kg or more.

☐ **What compound is formed by the binding of iron to deferoxamine?**

Ferrioxamine.

TOXICOLOGY AND ENVIRONMENTAL

☐ **Which types of iron preparations are least likely to be seen on abdominal x-ray after ingestion?**

Chewable and liquid formulations.

☐ **T/F: The deferoxamine challenge test is useful after iron overdose.**

False. The deferoxamine challenge involves administration of an IM deferoxamine dose. A positive test occurs if the urine changes color, indicating the presence of ferrioxamine. This test is considered unreliable.

☐ **T/F: Deferoxamine can chelate iron which is bound to ferritin and transferrin.**

False.

☐ **How do most children with acute, severe lead intoxication present to the ED?**

They are encephalopathic and often seizing.

☐ **What age group of patients are at greatest risk for lead toxicity?**

Children less than 36 months of age.

☐ **Is more ingested lead absorbed by the intestines in this population?**

Yes.

☐ **How is iron deficiency related to the possibility of lead toxicity?**

Iron deficiency for unknown reasons increases the intestinal absorption of lead.

☐ **Why may the brain of these patients be exposed to relatively more lead?**

They have an incomplete blood-brain barrier.

☐ **In addition to the above, why is childhood lead poisoning of greater concern than adult?**

More susceptible to lead's neurotoxic effects. Exposure of the developing brain can lead to permanent deficits in intelligence and behavior.

☐ **At what blood lead level is a child considered to be lead poisoned?**

10 mcg/dL.

☐ **Do children at this level need chelation therapy?**

No, but an investigation into possible lead sources and an emphasis on lead healthy lifestyle should ensue.

☐ **Is the number of lead poisoned children in the U.S. closer to 8% or 2%?**

8%.

☐ **Does lead (e.g. lead paint) have a noxious taste?**

No, it tastes sweet.

☐ **What are the principle sources of lead exposure?**

Lead in paint, soil, contaminated water, and dust.

☐ **Children of what racial group seems to be disproportionately affected?**

African American.

☐ **Can lead be inhaled?**

Yes, lead dust and lead oxide fumes are very efficiently absorbed by this route.

☐ **Is the unborn fetus at risk?**

Yes, lead readily crosses the placenta.

☐ **When did the lead content of paint begin to be regulated?**

1977.

☐ **What tests can be used in home inspections for lead?**

Sodium sulfide (produces lead sulfide, which is grayish black), and X-ray fluorescence.

☐ **Do folk, alternative, or herbal medications ever contain lead?**

Yes.

☐ **Which body compartment has the greatest lead content?**

Bone.

☐ **What X-ray finding is seen in growing children?**

Lead lines.

TOXICOLOGY AND ENVIRONMENTAL

☐ **Is it actual lead that is being imaged on these X-rays?**

No, they are growth arrest lines.

☐ **Which type of nerve is affected more by lead, sensory or motor?**

Motor. It is thought that lead neuropathy is an anterior horn cell disease with dying back of the motor neuron.

☐ **What characteristic finding is frequently seen on microscopic examination of the RBC in patients with prolonged lead poisoning?**

Basophilic stippling.

☐ **What two enzymes in the heme synthesis pathway are inhibited by lead?**

Delta aminolevulinic acid dehydratase and ferrochelatase.

☐ **What are two substances that accumulate due to this inhibition?**

Delta aminolevulinic acid (possibly neurotoxic), and erythrocyte protoporphyrin.

☐ **What is the utility of measure erythrocyte protoporphyrin in the blood?**

The effect on erythrocyte protoporphyrin lasts longer than elevated blood lead level, thus the effect of large exposure weeks earlier can be gauzed. It is not affected by "lead rebound," only by re-exposure.

☐ **Is the blood lead level a good indicator of total body burden of lead?**

No.

☐ **What are the two parenteral chelators available for severe lead poisoning?**

British Anti-Lewisite (BAL) and EDTA.

☐ **BAL cannot be given to patients allergic to what substance?**

Peanut oil, as this is its carrier.

☐ **In what group of patients must BAL be given cautiously?**

Patients with G6PD deficiency.

☐ **What substance should not be given orally to patients on BAL?**

Iron.

☐ **Should someone with encephalopathy and a blood level greater than 100 mcg/dL (70 mcg/L in children) be chelated parenterally?**

Yes.

☐ **Can EDTA be given orally?**

No. It is not well absorbed, and it may increase the absorption of lead.

☐ **What trace element deficiency can develop with EDTA chelation for lead?**

Zinc.

☐ **If a patient on EDTA has developed proteinuria, glycosuria, or hematuria, what has happened?**

Renal toxicity from the EDTA.

☐ **Does EDTA effectively decrease further a lead level 40 mcg/dL?**

No, probably not.

☐ **What are the oral and parenteral chelators that are available for lead chelation?**

Oral: Dimercaptosuccinic acid (DMSA, Succimer) and D-penicillamine.
Parenteral: EDTA and Dimercaprol.

☐ **What organ systems develop side effects of D-penicillamine?**

Gastrointestinal, dermatologic, bone marrow, and renal.

☐ **What are the side effects of DMSA?**

Rash, WBC depression, and LFT abnormalities. There will frequently be a rebound phenomena as lead from non-blood stores mobilize into the blood.

☐ **LSD is derived from what alkaloids?**

Ergots.

☐ **What is the onset of action and half-life of LSD?**

Onset of action is approximately 30 minutes after ingestion and peak effect is 3-5 hours. Elimination half-life is 2.5 hours.

☐ **What hallucinogenic alkaloid is found in Lophophora williamsii (peyote)?**

Mescaline.

TOXICOLOGY AND ENVIRONMENTAL

☐ **What neurotransmitter is responsible for the hallucinogenic action of LSD and the other hallucinogens?**

Hallucinogens are 5-HT2 receptor agonists.

☐ **What are the effects of LSD on the physical examination?**

Sympathomimetic symptoms predominate; tachycardia, tachypnea, mydriasis and diaphoresis.

☐ **What mushroom has hallucinogenic properties?**

Amanita muscaria.

☐ **What popular hallucinogenic drugs belong to the phenylethylamine class?**

Mescaline, MDMA (Ecstasy, 3,4 methylenedioxymethamphetamine), PMA (para-methoxyamphetamine), MDA (3,4-methylenedioxyamphetamine), methamphetamine and DOB (4-bromo-2,5-dimethoxyamphetamine).

☐ **Is the patient on a bad Lysergic Acid Diethylamide (LSD) "trip" oriented?**

Yes, even though they are fearful, and display bizarre reasoning.

☐ **Can the patient on LSD have dilated pupils and hyperthermia?**

Yes, dilated pupils are often seen, hyperthermia is seen in very high overdoses, presumably from the stimulation of 5-HT2 receptors.

☐ **Can the patient taking LSD have bruxism?**

Yes. Bruxism refers to constant chewing or biting motions characteristic of MDMA (Ecstasy) and LSD use/abuse, not often seen in other psychoactive medication use.

☐ **Is there an antidote for LSD?**

No, but benzodiazepines may be useful, often in large amounts.

☐ **What chemical is detected in the urine immunoassay for marijuana?**

11nor-delta 9-tetrahydrocannabinol-9-carboxylic acid (delta 9-THC-COOH).

☐ **What form of tetrahydrocannabinol is found in marijuana plant?**

The acid form, mostly acid A, of THC (delta 9 tetrahydrocannabinol) is found in the plant product. The THC acid A (and the acid B) must be heated to form THC.

PEDIATRIC EMERGENCY MEDICINE BOARD REVIEW

☐ **How are cannabinoids excreted?**

2/3 is excreted in the feces and 1/3 is excreted into the urine.

☐ **How long will a urine drug screen testing for delta 9-THC be positive after one exposure to marijuana?**

2-10 days.

☐ **How long will a urine drug screen testing for 11nor-9-carboxy-delta 9-THC be positive after chronic use of marijuana?**

Up to four weeks.

☐ **What is the main psychoactive chemical in marijuana?**

Delta-9-tetrahydrocannabinol (THC).

☐ **Is this chemical available legally?**

Yes, it is marketed as dronabinol (Marinol).

☐ **What is the mechanism of action of marijuana?**

It binds anandamide receptors in the brain. "Anandia" is translated from the Sanskrit as the word: "bliss".

☐ **What are the clinical symptoms of marijuana ingestion/overdose?**

It may have a variety of effects depending on a multitude of concurrent psychosocial or pharmacologic stimulation, but generally there is euphoria, a stimulant effect, and with higher doses a hallucinogenic and paranoid effect. There can also be a sedative effect and an orthostatic hypotensive effect after the initial stimulatory effect.

☐ **What are some signs of marijuana ingestion/overdose?**

Tachycardia, orthostatic hypotension, conjunctival injection, slurred sleep, incoordination, and stupor in excessive overdose.

☐ **Is there an antidote for a paranoid, psychotic patient overdosed on marijuana?**

There is no direct antidote; however, benzodiazepines would be effective. Orthostatic hypotension responds to fluids.

☐ **What is considered to be a lethal level of methemoglobin?**

70%.

TOXICOLOGY AND ENVIRONMENTAL

☐ **Pulse oximetry readings (oxygen saturation) approximate what value in the face of methemoglobinemia?**

85%.

☐ **At what level of methemoglobin can cyanosis be expected to be seen?**

15-20%.

☐ **What is the Valence State of iron in methemoglobinemia?**

+3 (ferric).

☐ **What mutated form of hemoglobin allows easier oxidation to methemoglobin?**

Hemoglobin M.

☐ **What is the reduced form of iron capable of oxygen transport?**

Fe^{2+} (ferrous).

☐ **What is methemoglobin?**

Hemoglobin in which the iron molecule has been oxidized to Fe^{3+} (ferric).

☐ **Why is there a baseline level of methemoglobin in the body?**

Oxygen transport occasionally results in a loss of an electron from the iron to the leaving oxygen molecule.

☐ **What concentration of methemoglobin is necessary to observe cyanosis in a patient?**

1.5 g/dL.

☐ **What concentration of sulfhemoglobin is necessary to observe cyanosis in a patient?**

0.5 g/dL.

☐ **What characteristic may one observes when drawing a blood sample from a patient with significant methemoglobinemia?**

"Chocolate brown" color of the blood.

☐ **Why is pulse oximetry abnormal in the patient with methemoglobinemia?**

The pulse oximetry device is designed to measure the absorption characteristics of oxyhemoglobin and deoxyhemoglobin. Methemoglobin has different absorption characteristics and will interfere with the readings.

PEDIATRIC EMERGENCY MEDICINE BOARD REVIEW

☐ **What pulse ox readings may one see in a patient with methemoglobinemia?**

As methemoglobin level rises, the pulse ox will drop until leveling off at 85% with methemoglobin concentrations of 30% or higher.

☐ **What type of device is needed to measure a methemoglobin level?**

A co-oximeter.

☐ **What is the significance of the "saturation gap" in a patient with methemoglobinemia?**

The patient's oxygen saturation measured by pulse oximetry will be lower than the calculated oxygen saturation on the arterial blood gas analysis. This is due to methemoglobin's interference with accurate pulse oximetry.

☐ **What substances may cause a falsely elevated methemoglobin level?**

Serum triglycerides, sulfhemoglobin and methylene blue.

☐ **What cause of methemoglobin has been associated with fatal cases in infants?**

Nitrates in well water.

☐ **What are the major classes of compounds that cause methemoglobinemia?**

Nitrites, nitrates, sulfonamides, quinines, bromates.

☐ **What is the baseline level of methemoglobin in a healthy child?**

Less than 1%.

☐ **What are the two enzymes that are responsible for reducing methemoglobin?**

NADH methemoglobin reductase and NADPH methemoglobin reductase.

☐ **What enzyme is responsible for the majority of methemoglobin reduction?**

NADH methemoglobin reductase.

☐ **Why may infants less than four months of age be more susceptible to methemoglobinemia?**

The enzyme NADH methemoglobin reductase does not reach full activity level until around this age.

☐ **What genetic defect may give patients a baseline elevation of methemoglobin up to 50%?**

NADH methemoglobin reductase deficiency.

TOXICOLOGY AND ENVIRONMENTAL

☐ **What vitamin may act to reduce methemoglobin?**

Ascorbic acid (vitamin C).

☐ **What is the antidote for methemoglobinemia?**

Methylene blue.

☐ **How does methylene blue help to reduce methemoglobin?**

Methylene blue is converted to leukomethylene blue by the NADPH methemoglobin reductase. The leukomethylene blue then reduces the methemoglobin.

☐ **At what level of methemoglobinemia may symptoms first appear?**

15%.

☐ **What are the first symptoms of significant methemoglobinemia?**

Dyspnea, fatigue, headache, dizziness.

☐ **What substances are associated with both methemoglobinemia and hemolysis?**

Amyl nitrite, aniline, dapsone, and phenazopyridine.

☐ **What is the normal rate of reduction of methemoglobin?**

15% per hour.

☐ **How is methylene blue administered in cases of methemoglobinemia?**

1-2 mg/kg of 1% solution IV over 5 minutes. A repeat dose of 1 mg/kg may be given.

☐ **What is the chemical name for methylene blue?**

Tetramethyl thionine chloride.

☐ **What is the major toxicity of dapsone overdose?**

Prolonged methemoglobinemia.

☐ **Why is methylene blue ineffective in patients with no G-6-PD activity?**

Patients with no G-6-PD are unable to generate NADPH in the red cell which is necessary for NADPH methemoglobin reductase to convert methylene blue to leukomethylene blue.

PEDIATRIC EMERGENCY MEDICINE BOARD REVIEW

☐ **What are the indications for methylene blue therapy?**

It is indicated in any patient with symptomatic methemoglobinemia.

☐ **What are possible side effects of large doses of methylene blue?**

Paradoxical methemoglobinemia and hemolysis.

☐ **Which local anesthetic agents may cause methemoglobinemia?**

Benzocaine and lidocaine.

☐ **What two analgesics may cause methemoglobinemia?**

Phenacetin and phenazopyridine.

☐ **Ingestions of which two types of mushrooms has been associated with renal failure?**

Cortinarius and Amanita.

☐ **Which group of mushrooms contains a chemical which can be hydrolyzed to monomethylhydrazine?**

Gyromitra species.

☐ **What percentage of mushroom-related deaths are caused by Amanita species?**

Approximately 90%.

☐ **What three types (genus) of mushroom contain cyclopeptides?**

Amanita, Galleria, and Lepiota.

☐ **When can Aminotransferase levels be expected to rise after Amanita poisoning?**

Within 2-3 days.

☐ **What endocrine abnormalities may be related to Amanita poisoning?**

Elevated insulin and calcitonin levels and depressed triiodothyronine levels.

☐ **Which cyclopeptide-containing mushroom produces a mixed polyneuropathy in 50% of cases?**

Lepiota species.

TOXICOLOGY AND ENVIRONMENTAL

☐ **What toxins do Amanita phalloides contain?**

Amatoxins, phallotoxins and virotoxins.

☐ **What is the mechanism of action of alpha-amanita?**

Inhibits RNA polymerase II, thereby preventing DNA transcription.

☐ **Is there any therapy for Amanita mushroom toxicity?**

Thioctic acid, penicillin, silibinin, cimetidine and multiple doses of activated charcoal have been suggested, but none have been definitively determined to be advantageous.

☐ **A patient has status epilepticus 6-10 hours after eating a mushroom that looked like a brain. Which mushroom did they eat?**

Monomethylhydrazine-containing mushrooms (Gyromitra).

☐ **What is the antidote for Gyromitra mushroom poisoning?**

Pyridoxine. Gyromitrin is metabolized to monomethylhydrazine which reacts with pyridoxine and inhibits GABA function leading to status epilepticus. Pyridoxine may limit this toxicity.

☐ **Which mushrooms cause a cholinergic toxidrome?**

Clitocybe and Inocybe.

☐ **What symptoms would you expect after a meal of Coprinus mushrooms and a glass of wine?**

Coprinus mushrooms contain the toxin coprine that can cause a disulfiram-like effect (hypotension, tachycardia, flushing, nausea and vomiting).

☐ **What is the toxicity of Amanita gemmate and Amanita muscaria?**

These mushrooms contain ibotenic acid and muscimol which produce a GABAergic syndrome of somnolence, hallucinations, myoclonus and can cause seizures.

☐ **What chemical toxin is contained in Conocybe cyanopus, Panaelus foenisecii, Gymnopilus spectabilis, Psilocybe caerulescens, and Psathyrella foenisecii?**

Psilocybin, a hallucinogenic compound.

☐ **Which class of mushroom causes delayed renal failure?**

Cortinarius species contain orellanine that can produce renal failure up to weeks after ingestion.

PEDIATRIC EMERGENCY MEDICINE BOARD REVIEW

☐ **What is lycoperdonosis?**

Lycoperdonosis is a syndrome of nasopharyngitis; vomiting, pneumonitis, shortness of breath; myalgias and fever caused by the inhalation of puffball mushrooms of the Lycoperdon species. Inhalation of spores is used as a treatment for epistaxis.

☐ **What is the toxic dose of amanitin?**

0.1 mg/kg.

☐ **How much Amanitin is contained in one Amanita species cap?**

10-15mg.

☐ **What is the theoretical mechanism of action for penicillin therapy for Amanita species toxicity?**

Penicillin prevents alpha-amanitin from entering the hepatocyte.

☐ **What is the mortality rate of patients who develop liver failure secondary to Amanita toxicity?**

20-30%.

☐ **What percentage of patients who call their Poison Center after ingesting a mushroom eventually develops symptoms?**

50%.

☐ **What percentage of patients who call their Poison Center after ingesting a mushroom eventually develop severe symptoms?**

0.2%.

☐ **If vomiting begins 30 minutes after ingestion, is that patient at increased risk of severe toxicity or decreased risk?**

If vomiting occurs prior to 3-4 hours after ingestion, the mushroom was most likely not toxic. However, many patients eat several mushrooms and may have eaten a mixture of toxic and non-toxic mushrooms.

☐ **What is the neonatal toxicity of Blue Cohosh?**

Myocardial depression and shock has occurred in neonates born to mothers who take Blue Cohosh.

☐ **Chronic ingestion of licorice can result in what electrolyte abnormality?**

Hypokalemia.

TOXICOLOGY AND ENVIRONMENTAL

☐ **Squill, oleander, foxglove and Ch'an Su result in toxicity identical to what medication?**

Digoxin. All of the above are cardiac glycosides.

☐ **Ingestion of jimson weed results in what toxidrome?**

Anticholinergic.

☐ **What is the active ingredient in St. John's Wort?**

Hypericin – it most likely acts as a serotonin reuptake inhibitor.

☐ **Which NSAID has been associated with agranulocytosis?**

Phenylbutazone.

☐ **NSAIDs inhibit what enzyme to produce their therapeutic effect?**

Cyclo-oxygenase.

☐ **What is the toxicity of massive ibuprofen overdose?**

Seizures, coma, metabolic acidosis, renal and liver failure.

☐ **What percentage of adverse drug reactions are related to NSAIDs?**

About 25%.

☐ **Which NSAID may cause a false-positive salicylate assay?**

Diflunisal.

☐ **Which veterinary NSAID has caused aplastic anemia and agranulocytosis?**

Phenylbutazone, formerly used in humans, can be highly toxic a relatively low dose (2g in a child; 4-40g in an adult) is now only available for veterinary medicine.

☐ **What is the unique aspect of NSAID toxicity secondary to the anthranilic acids (mefenamic acid and meclofenamate)?**

Reported rate of seizures in 20% of patients.

☐ **Serious manifestations of ibuprofen toxicity including coma, apnea and/or metabolic acidosis occur in what percentage of adults and children?**

9% of adults and 5% of children.

☐ **Which two NSAIDs will produce a false positive ferric chloride urine test (for salicylates)?**

Phenylbutazone and diflunisal.

☐ **What is the half-life of ibuprofen? Naproxen? Indomethacin?**

The half-lives of ibuprofen, naproxen and indomethacin are 2-4 hours, 10-20 hours and 3-11 hours, respectively.

☐ **What is the classic triad for opioid toxicity?**

Depressed mental status, miosis, and hypoventilation.

☐ **Which opioid may induce wide complex dysrhythmias?**

Propoxyphene.

☐ **Name three opioids more commonly associated with seizures.**

Meperidine, propoxyphene, tramadol.

☐ **What is the duration of action of naloxone?**

Approximately 1 hr.

☐ **Give the names of two opioids that have been associated with Serotonin Syndrome.**

Meperidine and dextromethorphan are both associated with serotonin syndrome.

☐ **How do opioids induce hypotension?**

Hypotension is mediated by histamine release. A combination of H1 and H2 antagonist is effective in improving hemodynamics effects of opioids in humans.

☐ **Which opioid in an overdose setting may result in an anticholinergic toxidrome?**

Diphenoxylate (Lomotil) is formulated with 0.025mg of Atropine.

☐ **What is the potentially harmful neurotoxic metabolite of meperidine?**

Normeperidine.

☐ **What is "Heroin Lung"?**

Noncardiogenic pulmonary edema, a complication of heroin overdose.

TOXICOLOGY AND ENVIRONMENTAL

☐ **What is the treatment of a narcotic overdose?**

Naloxone 0.4 to 2.0 mg IV over 2-3min maximum 10mg (0.1mg/kg in a child). If patient responds favorably consider IV infusion at 2/3 the effective dose.

☐ **What are the indications for use of naloxone?**

Respiratory depression (RR of 12 or less), coma and/or hypotension. Remember H1 and H2 blockers may also be effective in improving hypotension. Drowsiness alone is not an indication. In the setting of drowsiness, use of Narcan may result in withdrawal, and compromise of the airway, or aspiration pneumonia.

☐ **What effect does morphine have on the eye?**

It causes pupillary constriction (miotic).

☐ **What are the main pharmacological actions of morphine?**

Analgesia, respiratory depression, drowsiness and sedation, euphoria and tranquility, pupillary constriction, nausea and vomiting, and GI effects.

☐ **What type of analgesia does morphine produce?**

It produces analgesia without loss of consciousness, it is more effective against continuous dull pain than sharp intermittent pain, and it acts at both the spinal level and at higher levels.

☐ **What effect does morphine have on respiration?**

It produces respiratory depression primarily due to a decrease in sensitivity in the chemoreceptor centers to plasma CO_2.

☐ **Does morphine effect the cardiovascular system?**

It has mild effects there such as orthostatic hypotension.

☐ **How is morphine metabolized?**

It is mainly glucuronidated.

☐ **What are the side effects of morphine use?**

Dependence, increased pressure in the biliary tract, dysphoria and excitement, histamine release.

☐ **What type of withdrawal symptoms are seen with use of the opiates?**

It reflects an exaggerated rebound from the acute pharmacologic effects of the drug. The signs of withdrawal include rhinorrhea, lacrimation, yawning, chills, gooseflesh, hyperventilation, hyperthermia, mydriasis, muscular aches, vomiting, diarrhea, anxiety and hostility.

☐ **Name two drugs that are opiate antagonists?**

Naltrexone and naloxone.

☐ **What are the therapeutic uses of codeine?**

Mild to moderate pain and cough suppression.

☐ **What effect does meperidine have on the eyes?**

Unlike most opiates it causes the pupils to dilate rather than constrict. This is related to its anticholinergic activity.

☐ **What is diphenoxylate?**

A drug related to meperidine that is a potent anti-diarrheal agent.

☐ **What is loperamide?**

It is also related to meperidine and is a potent anti-diarrheal agent.

☐ **What is fentanyl?**

It is a synthetic opioid and a meperidine analog. Fentanyl and its derivatives including sufentanil are the most potent opioid agonists available.

☐ **How is fentanyl used?**

It is a potent analgesic. It is also used to produce neuroleptanalgesia and is used in cardiovascular surgery.

☐ **How is fentanyl administered?**

Either IV or transdermally or in lollipop form as preanesthetic medication for children.

☐ **How does naloxone work?**

It is an opioid receptor antagonist. It binds all opiate receptors to displace bound opioid agonists.

☐ **How does heroin differ from morphine?**

Heroin is hydrolyzed into morphine and therefore has many properties similar to morphine. It is however, more lipid soluble and can get across the blood brain barrier faster.

☐ **How is naloxone used therapeutically?**

To reverse the respiratory depression and coma of opioid overdose.

TOXICOLOGY AND ENVIRONMENTAL

☐ **What are the therapeutic uses of naltrexone?**

It is useful for treating opioid dependence because it has a longer duration of action than naloxone.

☐ **What are the adverse effects of naltrexone?**

Hepatotoxicity, nausea, sedation and headache.

☐ **What happens if naloxone or naltrexone is given in the absence of an opiate agonist?**

Usually there is no physiologic effect in this situation.

☐ **How do you treat an acute opioid overdose?**

Support respiration and use naloxone.

☐ **What is the usual cause of death in patients with organophosphate poisoning?**

Respiratory failure secondary to central depression of respiratory drive as well as weakness of the muscles of respiration.

☐ **Which group of pesticides forms generally reversible bonds with the cholinesterase enzyme?**

Carbamates.

☐ **Hexachlorocyclohexane is the chemical name for which pesticide?**

Lindane.

☐ **What are the two types of pyrethroid pesticides?**

Type I (do not contain a cyano group) and Type II (contain a cyano group).

☐ **What id the toxicity of Type II pyrethroids in humans?**

Paresthesia, nausea, vomiting, dizziness, seizures, and acute pulmonary edema.

☐ **How are the paresthesias secondary to Type II pyrethroids treated?**

Vitamin E.

☐ **Organophosphates, though less toxic to the environment, are _____ (more/less) toxic to humans acutely.**

More.

☐ **Approximately what percentage of homes in the U.S. uses pesticides?**

90%.

☐ **What are the routes of absorption of pesticides?**

Ingestion, inhalation, and/or skin absorption.

☐ **Which has a larger LD-50 in humans, malathion or parathion?**

Malathion (1,375 mg/kg) is "safer" than parathion (3 mg/kg).

☐ **What is the mechanism by which organophosphate poisonings cause toxicity?**

Irreversible phosphorylation (inhibition) of acetylcholinesterase.

☐ **What toxidrome does organophosphate and carbamate poisonings cause?**

Cholinergic.

☐ **What is the difference in the mechanism of action between organophosphates and carbamates?**

Carbamates are reversible.

☐ **The mnemonic 'DUMBELS' have been used to summarize the symptoms seen with excessive stimulation of which receptor subclass?**

Muscarinic.

☐ **To what symptoms do the letters of the mnemonic 'DUMBELS' refer?**

D-diarrhea, U-urination, B-bronchospasm/bronchorrhea, E-emesis, L-lacrimation, S-salivation.

☐ **Where are nicotinic cholinergic receptors found?**

Central Nervous System, autonomic ganglia, and the motor endplate.

☐ **Organophosphate poisoned patients may experience loss of consciousness, delirium, respiratory depression, and seizures, by the stimulation of which receptors in the CNS?**

Nicotinic receptors in the CNS.

☐ **Stimulation of which receptor in organophosphate poisoning results in muscle fasciculations?**

Nicotinic receptors at the motor endplate.

TOXICOLOGY AND ENVIRONMENTAL

☐ **In which group of patients may the typical cholinergic symptoms of acute organophosphate toxicity be absent?**

Children--often present only with neurobehavioral changes.

☐ **What is the "delayed neurotoxic syndrome" of organophosphate poisoning?**

Paresthesias, and weakness in the distal extremities.

☐ **When does this syndrome develop in relation to exposure?**

In 2-3 weeks.

☐ **What type of electromyography pattern does it generate?**

Denervation.

☐ **What is the "intermediate neurotoxic syndrome" of organophosphate poisoning?**

Weakness of the proximal limb muscles, muscles of respiration, and cranial nerve palsies.

☐ **When with regards to exposure does this syndrome present?**

In 1-4 days.

☐ **Can the weakness of the respiratory muscles be sufficiently severe to require mechanical ventilation?**

Yes.

☐ **What pharmacokinetics mechanism may account for the intermediate neurotoxic syndrome?**

Delayed absorption or long half-life.

☐ **Prolonged effect of succinylcholine as well as an exaggerated response to organophosphate poisoning may be seen in patients with a genetic deficiency of _____.**

Plasma cholinesterase (pseudocholinesterase).

☐ **What percentage of the population is deficient in this enzyme?**

3%.

☐ **Which of the two cholinesterases, plasma or RBC, is the preferred test in documenting exposure and monitoring resolution?**

RBC cholinesterase (true cholinesterase).

☐ **How long does it take for the action of this enzyme to return after it has been poisoned by organophosphates?**

60-90 days or 1/2 to 1% per day, i.e., the rate of new RBC release from the bone marrow.

☐ **Is there a clear relation in serum cholinesterase decrease and the severity of the symptoms of organophosphate poisoning?**

No.

☐ **What is usually the cause of death in untreated organophosphate poisoning?**

Increased respiratory secretions resulting in respiratory failure.

☐ **What is the first drug that should be given immediately in acute organophosphate poisoning?**

Atropine.

☐ **How much of the drug must be given?**

Sufficient to dry the airways of secretion.

☐ **What does pralidoxime and the other oximes do?**

They reversibly phosphorylate the acetylcholine receptors.

☐ **Within what time-frame must it be given in an organophosphate poisoning.**

Within 24-48 hours.

☐ **What are the possible side effects if given too rapidly.**

It should be given 1-2 g (20-50 mcg/kg in children) intravenously 30 min. It may cause tachycardia, laryngeal spasm, and muscle rigidity.

☐ **What is the time course of carbamate poisoning?**

Onset 15-20 min with resolution within 24 hours.

☐ **What is the effect of pralidoxime (2-PAM) in carbamate poisoned patients?**

It may enhance the acetylcholinesterase inactivation.

☐ **What is the most commonly used organochlorine pesticide in the U.S.?**

Lindane.

TOXICOLOGY AND ENVIRONMENTAL

☐ **What enzyme does lindane inhibit, causing seizures?**

Gamma Amino Butyric Acid (GABA).

☐ **What oral treatment may increase the excretion of organochlorines?**

Cholestyramine.

☐ **Synthetic pyrethroids, such as permethrin, are synthetic versions of pyrethrum, which is an oleoresin extract of _____.**

Chrysanthemum cinerariaefolium.

☐ **What are the symptoms of pyrethrin poisoning?**

Tremor, incoordination, salivation, diarrhea, and rarely death.

☐ **Paraquat is especially toxic to _____.**

Lungs.

☐ **What routine treatment given in ambulances and in emergency departments is contraindicated in paraquat poisoning?**

Oxygen.

☐ **Why?**

Generation of hydrogen peroxides and free radicals are increased.

☐ **What effect does paraquat have on skin and the gastrointestinal tract?**

Caustic.

☐ **What is the difference between diquat and paraquat?**

Less toxicity with diquat, not as concentrated in the lung.

☐ **What is the mechanism of toxicity from Dieffenbachia species?**

Dieffenbachia species contain calcium oxalate crystals and proteolytic enzymes that are packaged in raphides and bundled into idioblasts. When pressure is placed on the idioblasts, the crystals project outward and inject the enzymes under the skin or mucosa. The enzymes increase histamine and bradykinin release leading to local pain and swelling. Most exposures occur in small children who bite the leaves.

☐ **What is the toxicity of poinsettia (Euphorbia pulcherrima)?**

Poinsettia contains diterpene esters which may cause some local irritation or nausea and vomiting.

☐ **Are Philodendron species toxic?**

Philodendron species can rarely cause local or gastrointestinal irritation.

☐ **What is the laboratory abnormality associated with pokeweed (Phytolacca americana) toxicity?**

Leukocytosis. Phytolacca americana contains phytolaccine, a gastrointestinal irritant, and pokeweed mitogen, which causes lymphoid mitosis leading to leukocytosis on days 2-10. Toxicity is often avoided by double boiling (parboiling) the plant prior to ingestion.

☐ **What is the antigenic substance in the Toxicodendron species (poison ivy, poison oak and poison sumac)?**

Urushiol.

☐ **What is the toxicity of Datura stramonium (Jimson Weed)?**

Jimson Weed contains belladonna alkaloids, including atropine, scopolamine and hyoscyamine, that cause anticholinergic toxicity.

☐ **How much atropine is present in 100 seeds of Datura stramonium?**

100 Datura stramonium seeds contain the equivalent of 6mg of atropine.

☐ **What is the pattern of liver necrosis seen with Ackee fruit (Blighia sapida) toxicity?**

Centrilobular necrosis.

☐ **What is the toxic substance in water hemlock (Cicuta maculata)?**

Cicutoxin.

☐ **What is the antidote for Oleander toxicity?**

Oleander (Nerium oleander) contains cardiac glycosides and produce a digitalis-like toxicity. Digoxin-specific antibody fragments can be used to reverse toxicity, but particularly high doses may be necessary because of incomplete cross-reactivity.

☐ **Are Digoxin levels accurate in plant-based cardiac glycoside toxicity (e.g. ingestion of Lily of the valley)?**

Digoxin levels may be elevated in a patient with cardiac glycoside toxicity from ingestion of lily of the valley. However, digoxin levels may be lower than expected for the patient's toxicity because of incomplete cross-reactivity of naturally occurring glycosides and digoxin.

TOXICOLOGY AND ENVIRONMENTAL

☐ **What is the toxicity of Cicuta maculata (water hemlock)?**

Water hemlock causes nausea, vomiting, abdominal cramping and status epilepticus. The seizures may be refractory to standard therapy and is the probable cause of death in most cases.

☐ **Describe a death by poison hemlock (Conium maculatum).**

Conium maculatum contains coniine, a piperidine alkaloid similar to nicotine, that causes gastrointestinal upset, tachycardia, tremors, mydriasis, and seizures followed by bradycardia, and ascending muscle paralysis. Death is generally from paralysis.

☐ **What is the toxin contained in the Prunus species?**

Amygdalin.

☐ **What is the toxic chemical found in 'deadly nightshade'?**

Atropine and anticholinergic alkaloids.

☐ **What is the toxicity of Amygdalin?**

Amygdalin is found in the leaves or seeds of the Prunus species (e.g. apple seeds, apricot, cassava, cherry laurel, hydrangea, peach, plum and western chokecherry). Amygdalin is metabolized by the enzyme emulsin to form hydrocyanic acid. Toxicity is identical to cyanide toxicity.

☐ **What is the mechanism of action of Ricin?**

Ricin inhibits the 60S ribosomal subunit preventing binding of elongation factor 2 and inhibiting protein synthesis.

☐ **What is Melaleuca alternifolia?**

Melaleuca alternifolia is Tea Tree Oil and can cause sedation and ataxia.

☐ **What is the toxicity from Dogbane?**

Digoxin-like cardiac glycoside toxicity.

☐ **What is catnip?**

Catnip contains Nepeta cataria which is a mild hallucinogen.

☐ **What are the indications for hemodialysis in salicylate toxicity?**

Renal failure, Noncardiogenic pulmonary edema, Severe acid-base or electrolyte imbalance, despite appropriate and aggressive therapy. Coagulopathy secondary to hepatic failure. Salicylate level > 100 mg/dL (in acute overdose). Congestive heart failure (relative), Progressive deterioration in vital signs.

☐ **Oil of wintergreen contains what toxic substance?**

Oil of wintergreen contains 1.4 g/cc of salicylates.

☐ **Which type of acid base disturbance initially occurs with a salicylate overdose?**

Respiratory alkalosis. Approximately 12 hours later, an anion gap metabolic acidosis or mixed acid base picture may occur.

☐ **Is hyperglycemia or is hypoglycemia expected with a salicylate overdose?**

Expect either hyperglycemia or hypoglycemia.

☐ **A patient presents with an acute salicylate ingestion. What symptoms are expected with a mild, a moderate, and a severe overdose?**

1. Mild: Lethargy, vomiting, hyperventilation, and hyperthermia.
2. Moderate: Severe hyperventilation and compensated metabolic acidosis.
3. Severe: Coma, seizures, and uncompensated metabolic acidosis.

☐ **What is the treatment for a salicylate overdose?**

Decontaminate, lavage and charcoal, replace fluids, supplement with potassium, alkalize the urine with bicarbonate, cool for hyperthermia, administer glucose for hypoglycemia, place on oxygen and PEEP for pulmonary edema, prescribe multiple dose activated charcoal, and initiate dialysis.

☐ **What are the acid-base disturbances commonly present in salicylate overdose?**

Respiratory alkalosis and metabolic acidosis.

☐ **How do salicylates cause a respiratory alkalosis?**

Stimulation of the respiratory center of the brain.

☐ **How does the acid-base disturbance after salicylate overdose in pediatric patients differ from that in adults?**

In children the respiratory alkalosis is often transient and the metabolic acidosis more prominent.

☐ **What is the first acid-base change after salicylate overdose?**

The respiratory alkalosis.

☐ **What are the main metabolic effects of salicylate toxicity?**

Uncoupling of oxidative phosphorylation increased lactate production.

TOXICOLOGY AND ENVIRONMENTAL

☐ **What effect may salicylate toxicity have on the lungs?**

Non-cardiogenic pulmonary edema.

☐ **What liver disease in children has been associated with salicylate use?**

Reye's syndrome.

☐ **What risk factors have been associated with the development of pulmonary edema after salicylate overdose?**

Chronic overdose, neurologic symptoms, and smoking.

☐ **What neurologic effects may occur with salicylate toxicity?**

Tinnitus and hearing loss, vertigo, hyperventilation, agitation, and/or convulsions.

☐ **What is the pKa of salicylate?**

3.5.

☐ **What GI effects may be seen in salicylate toxicity?**

Nausea, vomiting, gastritis, and/or pylorospasm.

☐ **What is the therapeutic range for salicylate?**

15 to 30 mg/dL.

☐ **What is the normal protein binding of salicylate and how does it change after overdose?**

The normal protein binding of 90% falls to 75% with high serum levels.

☐ **What percentage of salicylate is excreted unchanged in the urine?**

2.5%.

☐ **What changes occur in the elimination of salicylate after overdose?**

Half-life is prolonged due to saturation of elimination pathways and change from first order to zero order kinetics.

☐ **Is the Done nomogram useful in predicting toxicity in salicylate overdoses?**

Rarely- the nomogram may be used only if very strict criteria are met including single overdose, non-enteric coated formulation, and serum pH 7.4 or higher.

☐ **Why does a low blood pH worsen salicylate toxicity?**

Low pH enhances salicylate concentration in body tissues such as the CNS.

☐ **Why does alkalinization of the urine enhance salicylate excretion?**

Alkalinization of the urine causes salicylate to become ionized in the urine, preventing re-absorption in the non-ionized state. This is referred to as "ion trapping".

☐ **What patients are at risk for chronic salicylate toxicity?**

The elderly and young children.

☐ **How do the symptoms of chronic salicylate toxicity differ from those in acute toxicity?**

Symptoms are similar, except that chronic toxicity most often manifests more pronounced neurologic symptoms and is more likely to cause pulmonary edema.

☐ **What qualitative bedside test may be used to detect salicylate?**

Addition of ferric chloride to urine will result in a dark purple color if salicylate is present.

☐ **What abnormality may be seen on urinalysis after salicylate overdose?**

Ketonuria and proteinuria.

☐ **Why is acetazolamide not indicated for alkalinizing the urine after salicylate overdose?**

Acetazolamide will increase the acidity of the blood and enhance distribution of salicylate to the body tissues.

☐ **What is the target urine pH level of the urine when alkalinizing to enhance salicylate elimination?**

7.5 to 8.

☐ **What electrolyte abnormality may make urinary alkalinization more difficult?**

Hypokalemia.

☐ **What is the treatment for severe salicylate toxicity?**

Hemodialysis.

☐ **What serum concentrate of salicylate is an indication for hemodialysis?**

100 mg/dL.

TOXICOLOGY AND ENVIRONMENTAL

☐ **What dose of aspirin is equivalent in salicylate to one cc of methyl salicylate?**

1.4 gm.

☐ **What is the active ingredient of oil of wintergreen?**

Methyl salicylate.

☐ **What is the half-life of a therapeutic dose of salicylate?**

2 to 4 hours.

☐ **Why may endotracheal intubation acutely worsen salicylate toxicity?**

During intubation, a patient who has been breathing rapidly to maintain a normal or near-normal pH may become acidotic due to decreased ventilation. Acidosis will enhance tissue distribution of salicylate.

☐ **Salicylate is structurally similar to what vitamin?**

Vitamin K.

☐ **What are the effects of salicylate toxicity on blood clotting?**

Hypoprothrombinemia and decreased platelet function.

☐ **What effect does theophylline have on cyclic AMP?**

Increases it.

☐ **What are the differences between theophylline and aminophylline?**

Aminophylline is the ethylenediamine salt of theophylline. Aminophylline may be given IV, theophylline by mouth only.

☐ **To what class of drugs does theophylline belong?**

Methylxanthines.

☐ **In what forms is theophylline used?**

Commonly as a sustained-release formulation, also as immediate release theophylline and aminophylline.

☐ **What are the likely mechanisms of action of theophylline?**

Phosphodiesterase inhibition, adenosine antagonism, catecholamine release.

☐ **What is the Vd of theophylline?**

0.5 L/kg.

☐ **What are the metabolic effects of acute theophylline toxicity?**

Hypokalemia, hypophosphatemia, hypomagnesemia, lactic acidosis, and hyperglycemia.

☐ **Above what serum theophylline level is serious acute toxicity likely to occur?**

90 to 100 ug/L.

☐ **Above what serum theophylline level is serious toxicity likely to occur in a chronic overdose?**

40 ug/mL.

☐ **What are the cardiovascular effects of acute theophylline toxicity?**

Sinus tachycardia, supraventricular tachycardia, atrial fibrillation, multifocal atrial tachycardia, hypotension, widened pulse pressure, and/or ventricular arrhythmias.

☐ **What are the CNS effects of acute theophylline toxicity?**

Anxiety, hyperventilation, tremor, seizure, status epilepticus.

☐ **What is the cause of hypotension in theophylline overdose?**

Beta-adrenergic mediated vasodilation.

☐ **What is the treatment for theophylline induced hypotension?**

IV fluids, Alpha-adrenergic agonists and Beta-adrenergic blockers.

☐ **What methods of enhanced elimination are proven to lower serum levels of the drug?**

Charcoal hemoperfusion, hemodialysis, and multi-dose activated charcoal.

☐ **What is the preferred method of enhanced elimination in severe theophylline toxicity?**

Charcoal hemoperfusion.

☐ **What is the clearance of theophylline by charcoal hemoperfusion?**

225 mL/min.

TOXICOLOGY AND ENVIRONMENTAL

☐ **What is the mechanism of theophylline-induced hypokalemia?**

Shift of potassium into cells via beta-adrenergic stimulation.

☐ **What percentage of theophylline is bound to plasma proteins?**

50%.

☐ **Name the three common methylxanthines.**

Theophylline, caffeine, theobromine.

☐ **Theophylline is an active metabolite of what drug?**

Caffeine - particularly in neonates.

☐ **Name four elimination products of theophylline.**

3-methylxanthine, 1,3-dimethyluric acid, 1-methylxanthine (which is further metabolized to 1-methyluric acid).

☐ **What is the active metabolite of theophylline?**

1,3-dimethyluric acid.

☐ **Why may orogastric lavage be ineffective in removing pills after a theophylline overdose?**

The pills are usually too large to fit into even large bore orogastric tubes.

☐ **What toxic effect of theophylline overdose may make therapy with activated charcoal difficult?**

Severe nausea and vomiting.

☐ **How often should activated charcoal be administered to a patient with severe theophylline toxicity?**

Every hour.

☐ **What are two possible adverse effect of charcoal administration in a sustained-release theophylline overdose?**

Vomiting and bezoar formation.

☐ **What is one possible problem with whole-bowel irrigation with PEG solution in a patient with sustained-release theophylline overdose?**

Decreased effectiveness of activated charcoal.

☐ **Should phenytoin be administered to a patient with theophylline-induced seizures?**

No.

☐ **How often should theophylline levels be checked after overdose?**

Every one to two hours.

☐ **What level of theophylline is an indication for charcoal hemoperfusion?**

90 ug/mL.

☐ **Name some indications for charcoal hemoperfusion in theophylline toxicity?**

Seizures, hypotension unresponsive to IV fluids, ventricular dysrhythmias, vomiting despite antiem

☐ **In what other instances may charcoal hemoperfusion be indicated?**

Rapidly rising theophylline level, significant underlying medical issues.

☐ **What abnormality may be seen on the CBC in theophylline overdose?**

Increased white blood cell count.

☐ **Which age groups of patients are at higher risk of toxicity in theophylline overdose?**

Neonates and the elderly.

☐ **What two mechanisms make activated charcoal effective in theophylline overdose?**

It binds the unabsorbed theophylline in the gut and decreases serum levels by "gastrointestinal dialysis".

☐ **What is the initial dose of fomepizole (4-MP) in ethylene glycol toxicity?**

15 mg/kg.

☐ **What is the toxic metabolite of methanol?**

Methanol is metabolized by alcohol dehydrogenase to formaldehyde which is metabolized by aldehyde dehydrogenase to formic acid. Formic acid causes the severe metabolic acidosis and blindness.

☐ **Which is the most intoxicating: methanol, ethylene glycol or isopropanol?**

CNS intoxication from alcohols is directly related to their number of carbon atoms. Isopropanol > ethylene glycol, ethanol > methanol.

TOXICOLOGY AND ENVIRONMENTAL

☐ **What are the indications for ethanol of fomepizole therapy in acute ethylene glycol ingestion?**

Ethylene glycol > 20mg/dL, osmol gap > 5 mosm/L and metabolic acidosis.

☐ **Which toxic alcohol ingestion is not effectively treated with 4-MP?**

Isopropyl.

☐ **What are the indications for hemodialysis in acute ethylene glycol ingestion?**

Ethylene glycol > 20 mg/dL, osmol gap > 10 mosm/L, metabolic acidosis and renal failure.

☐ **Which commercial products contain methanol?**

Windshield washing fluid, deicing solutions, carburetor cleaners, paint removers/thinners, model airplane fuels, antifreeze and gasohol.

☐ **How is methanol excreted?**

10% is excreted unchanged in the lungs. 3% is excreted unchanged in the kidneys and most is metabolized in the liver to CO_2 and H_2O.

☐ **Which commercial products contain ethylene glycol?**

Ethylene glycol is most commonly associated with antifreeze, but is also found in fire extinguishers, inks, pesticides, adhesives and in air conditioning and solar heating systems.

☐ **How is ethylene glycol excreted?**

Ethylene glycol is metabolized in the liver and 20% is excreted unchanged in the urine. This renal excretion accounts for all of the excretion of the drug in patients who are receiving ethanol or fomepizole.

☐ **What is the first metabolite of ethylene glycol?**

Ethylene glycol is metabolized by alcohol dehydrogenase to glycoaldehyde. Glycoaldehyde is metabolized to glycolic acid, then glyoxylic acid and then oxalic acid.

☐ **Which toxic alcohol is metabolized to lactic acid and maybe associated with both an osmolal gap and a severe anion gap metabolic acidosis?**

Propylene glycol.

☐ **What is turpentine made from?**

Pine oil.

☐ **What is "sudden sniffing death"?**

Inhalation of hydrocarbons for their euphoric effects, particularly halogenated hydrocarbons like trichloroethane, may sensitize the myocardium to the effects of catecholamines. Intense activity after "sniffing" or "huffing" may lead to ventricular fibrillation and sudden death.

☐ **What quality of hydrocarbon compounds describes the tendency of a liquid to become a gas?**

Volatility.

☐ **What quality of hydrocarbon compounds describes the tendency of a liquid to resist flow?**

Viscosity.

☐ **Hydrocarbon ingestions account for what percent of poison center calls in the U.S.?**

About 5%.

☐ **What is the primary target organ in serious petroleum distillate ingestions?**

The lungs.

☐ **Following ingestion of petroleum products, how do they enter the lungs?**

By aspiration, not via systemic spread.

☐ **What do aspirated hydrocarbons do in the lung?**

They interfere with surfactant activity.

☐ **What type of pneumonia results from the aspiration of the higher density hydrocarbons?**

Lipoid pneumonia.

☐ **What is the most worrisome cardiac effect following inhalation of halogenated hydrocarbons?**

Myocardial sensitization and lethal arrhythmias.

☐ **T/F: The findings on chest x-ray correlate well with symptoms following hydrocarbon aspiration.**

False.

☐ **What is the major toxicity of petroleum distillates?**

Pneumonitis.

TOXICOLOGY AND ENVIRONMENTAL

☐ **What characteristics (surface tension, viscosity, and volatility) of petroleum distillates increase the risk of aspiration?**

Low surface tension, low viscosity, and high volatility.

☐ **Which is the most important of these?**

Viscosity.

☐ **Do petroleum distillates arrive at the lung, where they are toxic, through gastrointestinal absorption (and transport via the blood stream) or through aspiration.**

Aspiration.

☐ **Do substances with viscosity >100 Saybolt Seconds Universal (SSU) have a high or low aspiration potential?**

Low.

☐ **What do petroleum distillates do to the lung when aspirated?**

Inhibit surfactant and also directly damage lung tissue.

☐ **Does the inhalation of petroleum distillates do the same?**

No.

☐ **What is the cardiac toxicity of hydrocarbons that are ingested?**

Sensitization to catecholamines, and resultant malignant arrhythmias.

☐ **Are there always chest radiograph abnormalities in significant hydrocarbon aspirations within the first few hours after ingestion?**

No.

☐ **If a patient after having ingested gasoline remains asymptomatic for 6 hours after ingestion, is it likely that the pulmonary toxicity will develop?**

No.

☐ **Is gastric decontamination (lavage, charcoal) recommended for petroleum distillates?**

No.

☐ **Why?**

Absorption is minimal, systemic toxicity is minimal, and decontamination increases the risk of aspiration.

☐ **Because terpenes (such as pine oil, or turpentine) are more viscous, they are less of an aspiration risk, but they have more toxic than other hydrocarbons to which two organ systems?**

CNS, GI.

☐ **All chlorinated short chain aliphatic hydrocarbons affect which two organ systems?**

CNS, skin.

☐ **What is "degreaser's flush"?**

Skin flushing seen when workers exposed to trichloroethylene drink ethanol.

☐ **What is the biochemical basis for this reaction?**

Competitive inhibition of aldehyde dehydrogenase.

☐ **What is the usual cause of death in acute poisoning from chlorinated hydrocarbons?**

Solvent induced narcosis.

☐ **Why is the liver toxicity of chlorinated hydrocarbons delayed by at least hours if not days?**

Toxicity is from bioreactive metabolites such as free radicals or epoxides.

☐ **A teenager presenting with severe generalized weakness, with the smell of paint/glue on him, has been abusing _____.**

Toluene.

☐ **How does this cause the severe weakness or paralysis?**

Toluene causes Type I (distal) Renal Tubular Acidosis and severe acute hypokalemia.

☐ **What is the main serious toxicity of camphor?**

Seizures.

☐ **Can 1 g (10 mL of Campho-Phenique) cause death in a child?**

Yes.

TOXICOLOGY AND ENVIRONMENTAL

☐ **This 10-year-old boy was playing with his mother's epinephrine auto-injector when it discharged into his finger. His thumb is erythematous and painful at the base, but pale and numb at the tip. He has no capillary refill in the tip of the finger. Is there a therapy for this?**

Phentolamine given subcutaneously may relieve the alpha vasoconstriction. Phentolamine was injected subcutaneously in this child's finger and capillary refill was regained in a few minutes. Warm compresses work best

☐ **What is the dose in an adult epinephrine auto-injector?**

Epinephrine auto-injectors contains 0.3 cc of 1:1000 epinephrine, or 0.3mg of epinephrine.

☐ **What is the possible toxicity related to a bite from this snake?**

Local toxicity will include swelling and pain. Systemic symptoms include nausea, vomiting, lightheadedness, metallic taste in mouth as well as thrombocytopenia, coagulopathy, renal failure, and hemolysis.

☐ **What are the toxicologic causes of basal ganglia infarct in children?**

Carbon monoxide and methanol.

☐ **You see this snake in North America. Is it poisonous?**

Yes. This snake can be identified as a pit viper by several characteristics: a pit located just below the halfway point between its eye and nostril, elliptical-shaped pupils, triangular head, and its rattle.

☐ **Is this snake's bite always poisonous?**

Rattlesnakes have about a 20% rate of 'dry bites', or bites that do not inject venom. Dry bites are more likely in defensive bites, rather than when the snake is hunting.

PEDIATRIC EMERGENCY MEDICINE BOARD REVIEW

☐ **Would the Crotalidae Polyvalent Anti-venin (Wyeth) treat this bite?**

Yes, the Crotalidae Polyvalent Anti-venin is active against rattlesnakes (crotalus and sistrurus), agkistrodon species (copperheads, water moccasins) and some South American and Asian snakes. This snake is a western diamondback.

☐ **How is the Crotalidae Polyvalent Anti-venin produced?**

The polyvalent anti-venin is produced with the venom of the western diamondback, eastern diamondback, tropical rattlesnake and the fer-de-lance. These venoms are injected into horses. The horse's serum is then used to isolate the IgG. Unfortunately, may patients have allergic reactions to this anti-venin and most patients who have several vials will develop serum sickness.

☐ **This patient has a carboxyhemoglobin level of 40% and is unconscious. What therapy is indicated?**

The patient should be placed on 100% oxygen as soon as possible. If the patient is at a center that has rapid access to a hyperbaric chamber, hyperbaric oxygen therapy may be helpful, particularly if the patient lost consciousness.

METABOLIC, ENDOCRINE AND NUTRITION

☐ **How does hyperglycemia lead to pseudohyponatremia?**

Because glucose stays in the extracellular fluid, hyperglycemia draws water out of the cell into the extracellular fluid. Each 100 mg/dL increase in plasma glucose decreases the serum sodium by 1.6 to 1.8 mEq/L.

☐ **What is the most common cause of hypovolemic hyponatremia in children?**

Viral gastroenteritis, with GI losses from vomiting and/or diarrhea.

☐ **What is the most common cause of euvolemic hyponatremia in children?**

Syndrome of inappropriate secretion of antidiuretic hormone (SIADH). Causes include CNS disorders, medications, tumors, and pulmonary disorders.

☐ **What are the signs and symptoms of hyponatremia?**

Weakness, nausea, anorexia, vomiting, confusion, lethargy, seizures, and coma.

☐ **What are the most common causes of hypotonic fluid loss leading to hypernatremia?**

Diarrhea, vomiting, hyperpyrexia, and excessive sweating, and iatrogenic.

☐ **What are the signs and symptoms of hypernatremia?**

Confusion, muscle irritability, seizures, respiratory paralysis, and coma.

☐ **What are the ECG findings of a patient with hypokalemia?**

Flattened T-waves, depressed ST segments, prominent P- and U-waves, and prolonged QT and PR intervals.

☐ **What are the ECG findings for a patient with hyperkalemia?**

Peaked T-waves: 5.5 – 6.6 mEq/L
P wave flattening, prolonged PR interval: 6.5 – 7.5 mEq/L
QRS prolongation: 7.0 – 8.0 mEq/L
Sine wave pattern: > 8.0 mEq/L

☐ **What is the first ECG finding for a patient with hyperkalemia?**

The development of tall peaked T-waves at levels of 5.6–6.0 mEq/L, which is best seen in the precordial leads.

PEDIATRIC EMERGENCY MEDICINE BOARD REVIEW

☐ **What is the quickest way to stabilize the myocardium in a patient with hyperkalemia?**

Administration of calcium gluconate (10%) 10-20 mL, IV with an onset of action of 1-3 minutes.

☐ **What are some causes of hyperkalemia?**

Acidosis, tissue necrosis, hemolysis, blood transfusions, GI bleed, renal failure, Addison's disease, primary hypoaldosteronism, excess po K+ intake, RTA IV, and medication (such as succinylcholine, beta-blockers, captopril, spironolactone, triamterene, amiloride, and high dose penicillin).

☐ **What are the causes of hypocalcemia?**

Shock, sepsis, multiple blood transfusions, hypoparathyroidism, vitamin D deficiency, pancreatitis, hypomagnesemia, alkalosis, phosphate overload, chronic renal failure, hypoalbuminemia, and medication, such as Dilantin, phenobarbital, heparin, theophylline, cimetidine, and gentamicin.

☐ **What is the most common cause of hyperkalemia?**

Lab error. Chronic renal failure is the most common cause of "true hyperkalemia."

☐ **What are the most common causes of hypercalcemia?**

In descending order, malignancy, primary hyperparathyroidism, and thiazide diuretics.

☐ **What are the signs and symptoms of hypercalcemia?**

Stones:	renal calculi.
Bones:	osteolysis.
Abdominal groans:	anorexia, constipation, peptic ulcer disease and pancreatitis.
Psychic overtones:	psychiatric disorders.

☐ **What is the initial treatment for hypercalcemia?**

Patients with hypercalcemia are dehydrated because high calcium levels interfere with ADH and the ability of the kidney to concentrate urine. Therefore, the initial treatment is restoration of the extracellular fluid with 2 times maintenance level of normal saline within 24 hours.

☐ **What is the most common cause of hyperphosphatemia?**

Acute and chronic renal failure.

☐ **What are the two primary causes of primary adrenal insufficiency?**

Tuberculosis and autoimmune destruction account for 90% of the cases.

METABOLIC, ENDOCRINE AND NUTRITION

☐ **What are the signs and symptoms of primary adrenal insufficiency?**

Fatigue, weakness, weight loss, anorexia, hyperpigmentation, nausea, vomiting, abdominal pain, diarrhea, salt craving, and orthostatic hypotension.

☐ **What characteristic lab findings are associated with primary adrenal insufficiency?**

Hyperkalemia, hyponatremia, hypoglycemia, azotemia (if volume depletion is present), and a mild metabolic acidosis.

☐ **How should acute adrenal insufficiency be treated?**

Administration of hydrocortisone (2 mg/kg) and crystalloid fluids containing dextrose.

☐ **What are the main causes of death during an adrenal crisis?**

Circulatory collapse marked by refractory hypotension, and hyperkalemia-induced arrhythmias.

☐ **What are the causes of acute adrenal crisis?**

It occurs secondary to a major stress, such as surgery, severe injury, or any other illness in a patient with primary or secondary adrenal insufficiency.

☐ **What is thyrotoxicosis, and what are the causes?**

A hypermetabolic state that occurs secondary to excess circulating thyroid hormone caused by thyroid hormone overdose, thyroid hyperfunction, or thyroid inflammation.

☐ **What is the most common cause of hypoglycemia seen in the ED.**

An insulin reaction in a diabetic patient.

☐ **What is the role of phosphate replacement during the treatment of DKA?**

Phosphate is gently recommended for a serum phosphate level <1.0 mg/L.

☐ **What is the most important "initial" step for the treatment of DKA?**

Fluid administration with 20 cc/kg normal saline.

☐ **In the first two years of life, what is the most common cause of drug-induced hypoglycemia?**

Salicylates. Between 2 and 8, alcohol is the most likely cause, and between 11 and 30, insulin and sulfonylureas are the culprits.

☐ **What is the most common type of hypoglycemia in a child?**

Ketotic hypoglycemia. Attacks usually occur when the child is stressed by caloric deprivation. This condition usually develops in boys between 18 months and 5 years of age. Attacks may be episodic and are more frequent in the morning or during periods of illness.

☐ **What are the neurologic signs and symptoms of hypoglycemia?**

Hypoglycemia may produce mental and neurologic dysfunction. Neurologic manifestations can include paresthesias, cranial nerve palsies, transient hemiplegia, diplopia, decerebrate posturing, and clonus.

☐ **What lab findings are expected with diabetic ketoacidosis?**

Elevated ß-hydroxybutyrate, acetoacetate, acetone, and glucose. Ketonuria and glucosuria are present. Serum bicarbonate levels, pCO2, and pH are decreased. Potassium may be initially elevated but falls when the acidosis is corrected.

☐ **What is the most common cause of secondary adrenal insufficiency and adrenal crisis?**

Iatrogenic adrenal suppression from prolonged steroid use. Rapid withdrawal of steroids may lead to collapse and death.

☐ **Outline the basic treatment for DKA.**

Give insulin via an IV drip at 0.1 units/kg/hr. Hydrate with normal saline at 20 cc/kg bolus (repeat times 2)

☐ **An increase of pCO2 of 10 mm Hg leads to an expected decrease in pH of about:**

0.08.

☐ **A decrease of pCO2 of 10 mm Hg leads to an expected increase in pH of about:**

0.13.

☐ **What expected increase in pH is associated with a rise in HCO3 of 5.0 mEq/L?**

0.08.

☐ **What expected decrease in pH is associated with a decrease in HCO3 of 5.0 mEq/L?**

0.10.

☐ **How is the anion gap calculated from electrolyte values?**

Anion gap = Na–Cl–CO2. The normal gap is 12 +/- 4 mEq/L.

METABOLIC, ENDOCRINE AND NUTRITION

☐ **What is the most common endocrine disorder of childhood and adolescence?**

Diabetes mellitus (DM).

☐ **What are the two primary causes of metabolic alkalosis?**

Loss of hydrogen and chloride from the stomach and overzealous diuresis with loss of hydrogen, potassium, and chloride.

☐ **What are the peak age groups in children who develop IDDM?**

Between the ages of 5 and 7 and at puberty.

☐ **What is the equation for determining serum osmolarity?**

Osmserum = [serum Na+(mEq/L) + K+(mEq/L)] x 2 + (glucose(mg/dL) / (18) + (BUN(mg/dL) / (3).

☐ **What condition should be suspected in a child who presents with lethargy, weight loss, and new onset enuresis in a previously toilet trained child?**

New onset DM.

☐ **How does a child with DKA present?**

The child will present with vomiting, polyuria, dehydration, Kussmaul breathing and abdominal pain.

☐ **What are the dangers of administering HCO3- to a child with DKA?**

The dangers of bicarb use in DKA:
Shift's O2 dissociation curve left causing decreased O2 release to tissues.
Causes paradoxical CSF acidosis.
Causes iatrogenic hypokalemia.
Alkalosis leading to dysrhythmias.
Delays fall in serum ketones.
Patients may overall do worse.

☐ **What is the Primary life threatening complication in the aggressive treatment of DKA?**

Cerebral edema, rare in children > 5 years. It occurs in < 1% of pediatric patients in DKA. Often in patients with more severe diseases. It carries a 40% - 90% mortality and is responsible for 50% - 60% of deaths due to DKA.

☐ **What signs and symptoms should a physician look for when considering cerebral edema in a patient being treated for DKA?**

Change in mental status, "delirious outbursts", bradycardia, vomiting, decreased reflexes, and changes in pupillary response.

PEDIATRIC EMERGENCY MEDICINE BOARD REVIEW

☐ **What percent of pediatric patients have DKA at their first hospital presentation?**

DKA is present at 25% - 40% of initial presentations of newly diagnosed juvenile onset diabetics.

☐ **What is the most serious immediate life threatening risk to a child with DKA?**

Dehydration. While patients may have hyperglycemia and metabolic acidosis, dehydration is the most serious life threatening condition.

☐ **When a child in DKA presents to the emergency department, The K+ will be increased, decreased, or normal?**

Total body K+ is depleted though serum levels may be normal. Transcellular shifts during acidosis move K+ extracellular in exchange for H+ ions. The osmotic diuresis due to the hyperglycemia results in overall K+ loss. In addition, the dehydration stimulates aldosterone secretion which further contributes to K+ excretion.

☐ **What are the risk factors for development of cerebral edema in a patient in DKA?**

High CO_2- and BUN at presentation, and HCO_3- use.

☐ **How do you calculate the water deficit in a patient with DKA?**

0.10 x kg (patient's wgt) The fluid deficit is estimated to be approximately 10% of the patient's weight in kilograms.

☐ **How do you determine if a child in DKA requires HCO_3-?**

It should not be used unless the pH is < 7.0. Acidosis will be corrected with fluids and insulin.

☐ **When should glucose be added to the IVF in the treatment of DKA?**

When the blood glucose level reaches 250 mg/dL the IVF should be changed to D5.45%NS.

☐ **How often should you monitor the serum K+ in a child with DKA?**

The K+ should be checked every 2 - 4 hours until acidosis is corrected.

☐ **Why do DKA patients experience changes in vision following the initiation of therapy?**

Rehydration in conjunction with the correction of hyperglycemia causes fluid shifts in the eyes with reversible lens distortion.

☐ **What are the signs and symptoms of hypoglycemia?**

They are nonspecific but may include seizures, tremulousness, confusion, palpitations, sweating, and irritability.

METABOLIC, ENDOCRINE AND NUTRITION

☐ **How is a child with hypoglycemia treated?**

Administer 2.5 ml/kg of 10% dextrose. If convulsions are present increase to 4ml/kg of 10% dextrose. Maintain the blood glucose level using 8 mg/kg/min of a 10% dextrose solution.

☐ **What is the most common cause for hypothyroidism in the neonate?**

Thyroid dysgenesis.

☐ **What is the most common thyroid anomaly in childhood?**

The thyroglossal duct cyst which is remnant of the thyroglossal duct.

☐ **Are infants born with complete thyroid agenesis symptomatic at birth?**

No. Transplacental T4 provides adequate thyroid hormone however the TSH level will still be elevated enough for detection of the abnormality.

☐ **A mass at the base of the tongue in a newborn with adequate T4 levels is suggestive of what disorder?**

A sublingual thyroid or a thyroglossal duct cyst.

☐ **What is the possible cause for a transient congenital hypothyroidism?**

Maternal antibodies that inhibit TSH binding.

☐ **What are the signs of neonatal hypothyroidism?**

Poor appetite, sluggishness, macroglossia, somnolence, large abdomen, hypothermia, molted skin, constipation, edema and bradycardia.

☐ **What is the treatment for neonatal hypothyroidism?**

Sodium-L-Thyroxine (10 - 15 mcgs/kg in neonates).

☐ **What is the most common cause of a acquired hypothyroidism?**

Lymphocytic thyroiditis.

☐ **What is the most common clinical manifestation of lymphocytic thyroiditis?**

The appearance of a painless goiter.

☐ **What is the differential diagnosis of a congenital goiter?**

Antithyroid drugs or iodine containing medications used during pregnancy, hyperthyroid infants, iodine deficient infants and rarely congenital teratoma.

PEDIATRIC EMERGENCY MEDICINE BOARD REVIEW

☐ **Is Grave's disease more common in boys or girls?**

Girls are affected five times as much as boys.

☐ **What are the classical signs and symptoms of Grave's disease?**

Emotional lability, autonomic hyperactivity, exophthalmos, weight loss or increased appetite with no weight gain, diarrhea, and a goiter in nearly all affected individuals. In pediatrics, look for anxious appearing, fidgety movements with a tremor.

☐ **How is Grave's disease confirmed by laboratory testing?**

One would find increased bound and free T3 and T4 with decreased TSH.

☐ **What treatment options exist for children with Grave's disease? What treatment is recommended?**

Children can be managed medically with propylthiouracil(PTU) or methimazole, surgically with a subtotal thyroidectomy or with radioiodine therapy. Medical management is the treatment of choice.

☐ **What are the differences between PTU and methimazole?**

Methimazole is ten times more potent on a weight basis allowing it to be a once a day medication. PTU is predominately protein bound making it a better choice in pregnant and nursing mothers as its ability to cross the placental membrane is limited.

☐ **What are the respective doses of PTU and methimazole?**

PTU is given at 5 - 10 mg/kg/d dived T.I.D. while methimazole is .5 - 1.0 mg/kg/d given either QD. or B.I.D.

☐ **What is the most common physical exam finding in Grave's disease?**

Goiter.

☐ **What medication is available to treat the autonomic hyperactivity associated with Grave's disease?**

Propranolol, a beta blocker dosed at .5 - 2.0 mg/kg/d divided T.I.D.

☐ **What is the most common cause for congenital hyperthyroidism?**

Transplacental passage of thyroid receptor autoantibodies from a mother with Grave's disease.

☐ **What are the three largest concerns with surgical management of Grave's disease?**

Hyper or hypothyroidism depending on the amount of tissue removed, hypoparathyroidism, and vocal cord paralysis.

METABOLIC, ENDOCRINE AND NUTRITION

☐ **What is the most common disorder of thyroid gland function in children presenting to the ED?**

Thyrotoxicosis.

☐ **What commonly occurs in the newborn whose mother suffered from hyperparathyroidism during pregnancy?**

Hypocalcemia resulting from suppression of the fetal parathyroids due to the elevated serum calcium in the maternal circulation.

☐ **What is the most common disorder causing thyrotoxicosis in children?**

Grave's disease.

☐ **What percent of children with hyperthyroidism experience a thyroid storm?**

Approximately 1%.

☐ **What are the symptoms of child suffering from a thyroid storm?**

Tachycardia, systolic hypertension, tremulousness, delirium and hyperthermia.

☐ **How do you manage acute thyrotoxicosis?**

Propranolol at 10 mg/kg IV over 10 - 15 minutes for hypertension and increased metabolic rate. Lugol's iodide, 5 drops PO every eight hrs or Na+ iodide 125 - 250 mg/d IV over 24 hrs will stop thyroxine production.

☐ **What measures can be taken to further reduce peripheral conversion of T4 to T3?**

Oral dexamethasone at .2 mg/kg or oral hydrocortisone 5 mg/kg.

☐ **What is the differential diagnosis of acute primary adrenal insufficiency?**

Congenital adrenal hypoplasia, autoimmunity, TB, infection, trauma, and adrenal hemorrhage.

☐ **What are the clinical manifestations of adrenocortical insufficiency?**

Low blood pressure, muscular weakness, weight loss, anorexia and salt craving.

☐ **What emergency should be suspected in patients who present with cyanosis, cold skin, thready pulse, hypotension and tachypnea?**

Addison's crisis.

☐ **What conditions have been implicated in precipitating adrenal crisis?**

Infection, trauma, fatigue, and various medications.

PEDIATRIC EMERGENCY MEDICINE BOARD REVIEW

☐ **Which symptom is most commonly seen in children with Addison's disease?**

Hypoglycemia.

☐ **What is the Waterhouse-Friderichsen Syndrome?**

Primary adrenal insufficiency due to adrenal hemorrhage. It is often secondary to meningococcemia induced shock.

☐ **What are the causes of secondary adrenal insufficiency?**

Diminished ACTH levels secondary to long term glucocorticoid treatment, traumatic brain injury, and sepsis.

☐ **At what decreased Na+ concentration do children generally become symptomatic?**

When the serum Na+ is less than 120 mEq/L.

☐ **What syndrome is likely to be present in a 7-year-old child who presents with nausea, vomiting, and a single seizure with laboratory results consistent with hyponatremia and increased urine osmolarity?**

The syndrome of inappropriate antidiuretic hormone (SIADH).

☐ **What is SIADH?**

It is a the syndrome of inappropriate ADH(antidiuretic hormone) secretion. It occurs when the plasma vasopressin levels are abnormally elevated for the corresponding plasma osmolarity.

☐ **What is the most common cause of SIADH in children?**

Infection. 50% of children with bacterial meningitis have concomitant SIADH.

☐ **How is SIADH treated?**

Most important is treating the underlying condition. Fluid restriction initially. In emergent situations, furosemide, in conjunction with 300 ml/M2 of 1.5% NaCl may be used. Side effects and nephrotoxicity limit the use of demeclocycline and lithium.

☐ **What is the emergent treatment for a hyponatremic child suffering from SIADH?**

3 ml/kg of 3% saline q10-20 minutes until the symptoms resolve. A single dose of 1 mg/kg of furosemide may be used. 5 - 10 mg/kg of phenytoin IV can be used to inhibit ADH secretion as well as help to prophylax against possible seizures. Treating the underlying condition however is still the most important consideration.

☐ **What is the danger of correcting the Na+ too vigorously?**

Central pontine myelinolysis.

METABOLIC, ENDOCRINE AND NUTRITION

☐ **How is the diagnosis of diabetes insipidus made in a child?**

By finding hypernatremia, increased serum osmolarity and decreased urine osmolarity.

☐ **What is the danger of vigorously hydrating a patient who has hypernatremia due to diabetes insipidus?**

Cerebral edema, seizures and death.

☐ **What additional adjuncts can be used in addition to volume replacement in a patient with central diabetes insipidus?**

Intranasal administration of DDAVP at 0.4 g/kg.

☐ **Is Cushing's disease different from Cushing's syndrome?**

Cushing's Disease: Pituitary adenoma
Cushing's syndrome: High cortisol from paraneoplastic, adrenal tumor, iatrogenic.

☐ **What is the most common cause of excess cortisol production in infants?**

A functioning adrenocortical tumor.

☐ **What is the most common cause of childhood hyperparathyroidism?**

A single parathyroid adenoma.

☐ **What are the clinical manifestations of hyperparathyroidism regardless of cause?**

Anorexia, nausea, vomiting, constipation, polydipsia, polyuria, fever, weight loss, kidney stones and muscular weakness.

☐ **What is the most common cause for secondary hyperparathyroidism in children?**

Chronic renal disease. The formation of the active form of vitamin D takes place in the kidney. With diseased kidneys the active form of vitamin D is not made resulting in decreased intestinal absorption of calcium.

☐ **What is the treatment of choice for symptomatic hypercalcemia (>15 mg/dL)?**

Isotonic saline bolus is the first treatment, followed by lower rate continuous infusion, and close monitoring.

☐ **In a child with resolving DKA, when do you cease insulin infusion?**

When ketones have been cleared, metabolic acidosis has resolved, blood glucose has normalized and the patient can eat and drink.

PEDIATRIC EMERGENCY MEDICINE BOARD REVIEW

☐ **A child is brought in obtunded and with acute hypoglycemia, what should be the first treatment modality?**

IV glucose infusion (2-3 mg/kg of D10W or 1ml/kg of D25W).

☐ **What causes pseudohyponatremia?**

Hyperglycemia, severe hypertriglycemia, hyperproteinemia, lab error.

☐ **What are the daily sodium requirements of premature and full term newborns?**

3-4 mEq/kg/d for premature and 1-2 mEq/kg/d for full term.

☐ **In which conditions are electrolytes considered high yield?**

Age <6 months, vomiting, tachycardia, dry mucous membrane, and capillary refill >2 seconds. The presence any of these criteria is 100% sensitive for significant abnormalities.

☐ **What is the formula for obtaining the true calcium level with hypoalbuminemia?**

Subtract 0.8 mg/dL from total serum calcium for each 1 g/dL decrease in serum albumin below normal.

☐ **What are the clinical findings in patient with hypocalcemia?**

Milder signs: muscle cramps, perioral or finger paresthesias, SOB from bronchospasm.
Severe signs: hypotension, long QT, angina, CHF, dysrhythmias.
Chronic signs: Chvostek's and Trousseau's signs.

☐ **What is the relationship between acidosis and hyperkalemia?**

A decrease in the pH of 0.1 will increase K+ 0.3-1.3 mEq/L due to the H+/K+ exchange pump.

SUDDEN INFANT DEATH SYNDROME AND APPARENT LIFE-THREATENING EVENT

☐ **Sudden Infant Death Syndrome (SIDS) is the most common cause of death for infants between the ages of 1 month to 1 year. It occurs at a rate of 2500/year. What are 6 risk factors associated with SIDS?**

1. Prematurity with low birth weight.
2. Previous episode of apnea or apparent life-threatening event (ALTE).
3. Mother is a substance abuser.
4. Family history of SIDS.
5. Male gender.
6. Low socioeconomic status.

☐ **What are some of the modifiable risk factors for sudden infant death syndrome (SIDS).**

Prone sleeping position ("Back to Sleep"), maternal smoking, lying infant on soft bedding (use firm flat sleep surface), breast feed infant, don't sleep in the same bed, using a pacifier during sleep decreases the incidence of SIDS, keep the ambient temperature comfortable, not too warm.

☐ **SIDS has a bimodal distribution. At what ages do the peaks occur?**

2 and 4 months of age. 90% of cases occur before 6 months.

☐ **Define failure to thrive (FTT).**

FTT is defined as an infant who is below the third percentile in height or weight or whose weight is less than 80% of the ideal weight for their age. Almost all patients with FTT are under 5 years old, while the majority of children are 6–12 months old.

☐ **What is the most common cause of FTT?**

Poor intake is responsible for 70% of FTT cases. One third of these cases are educational problems, ranging from inaccurate knowledge of what to feed a child to over-diluting formula to "make it stretch further."

☐ **What is the prognosis for patients with FTT?**

Only 1/3 of patients with FTT due to environmental factors have a normal life. The remainder grow up small for their size, and the majority also have developmental, psychological, and educational deficiencies.

☐ **What is the definition of apnea in an infant?**

Apnea is either cessation of breathing for more than 20 seconds, or a shorter period associated with cyanosis or bradycardia.

PEDIATRIC EMERGENCY MEDICINE BOARD REVIEW

☐ **What is the definition of ALTE?**

An Apparent Life Threatening Event is apnea associated with color change (cyanosis or redness), a loss of muscle tone, and choking or gagging.

☐ **The term ALTE has recently been changed. What is the new terminology?**

BRUE or Brief Resolved Unexplained Event. ALTE implied that something potentially serious had happened to the infant and required a workup and/or admission. The American Academy of Pediatrics (AAP) clinical practice guidelines in 2016 changed the term.

☐ **What is the definition of BRUE?**

An event occurring in an infant < 1 year when the observer reports a sudden, brief, and now resolved episode of > 1 of the following: 1) cyanosis or pallor; 2) absent, decreased, or irregular breathing; 3) marked change in tone (hyper/hypotonia); 4) altered level of responsiveness and no explanation after a thorough, appropriate history and physical exam.

☐ **Who is considered a low risk infant from a BRUE?**

Age > 60 days, gestational age > 32 weeks, one event only (no prior BRUE, no cluster of BRUEs), the BRUE lasted < 1 minute, the event did not require CPR, the history and physical are normal.

☐ **Who is considered a high risk infant from a BRUE?**

Infants not meeting age or timing criteria; abnormal history or physical exam; family history of sudden cardiac death, or nondiagnostic social, feeding or respiratory problems.

☐ **What is the emergency department intervention for a low risk infant from a BRUE?**

The main intervention is to educate the parents about BRUE. Assure parents that their child is at a low risk and the infant is unlikely to have any adverse outcomes. Obtain an ECG, place on a pulse oximeter and cardiac monitor and repeat an exam during the monitoring process. There is no specific recommendations regarding the duration of monotoring. Labs, chest x-ray, echo, reflux or seizure medication, or admission for observation is not required.

☐ **How often is a definitive diagnosis found in a patient with a BRUE?**

50%.

☐ **Compared to an otherwise healthy infant, how much more likely is a child with a history of a BRUE to go on to have SIDS?**

About 3-5 times more likely

☐ **What percentage of infants who have SIDS had a prior BRUE?**

About 5%

Neurology

☐ **What are causes that predispose a child to arterial ischemic strokes?**

Prematurity, Sickle Cell anemia, cyanotic or acquired congenital heart disease, thrombophilias, sepsis with or without meningitis, varicella, trauma and arterial venous malformation.

☐ **What data exist about the use of TPA for the treatment of stroke in neonates, infants and children?**

There is little data regarding the use of TPA for the treatment of strokes in patients less than 18 years of age. However, thrombolytic drugs may have some role in very select and unusual pediatric clinical entities.

☐ **What are signs of increased intracranial pressure?**

Headache is the ealiest symptom of increased ICP. Altered mental status (combativeness, irritability, somnolence, inconsolable), abnormal pupils (pupillary dilatation, lack of response to light, unequal size), nausea, vomiting, abnormal gait and abnormal posturing. Also, an increased head circumference and bulging fontanelles may be seen on exam.

☐ **What are indications for intubation of a comatose child?**

Depressed mental status can be associated with hypoventilation (and the development of hypoxia and hypercarbia), loss of protective air reflexes and increased intracranial pressure. Intubation is indicated for oxygenation, control of intracranial hypertension (through mild hyperventilation) and protection of the airway.

☐ **What is Cushing's triad?**

Cushing's triad is the combination of bradycardia, hypertension and irregular ineffective respiration. It is a late sign of increased intracranial pressure and impending herniation.

☐ **What bacterial illnesses present with peripheral neurologic findings?**

Botulism, tetanus, diphtheria, Brucella (peripheral and cranial nerve neuropathies), and Campylobacter (associated with Guillian-Barre Syndrome).

☐ **A child with a history of strabismus complains of blurry vision. On exam she has an abnormal pupil reflex and a white reflex on funduscopic examination. What is the treatment?**

Retinoblastoma. This tumor can spread to other sites in the brain or body. In the past surgical removal of the eye was the only option. There are now vision-sparing options depending on the classification. First line therapy continues to be enucleation.

PEDIATRIC EMERGENCY MEDICINE BOARD REVIEW

☐ **What area of the brain is dysfunctional when a patient has Cheyne-Stokes respirations?**

The cortex. The nervous system is relying on diencephalic control.

☐ **What will happen if you shine a light in the eyes of a patient who is in a diabetic coma?**

The pupils will constrict.

☐ **A 15-year-old presents with a history of being knocked unconscious for 10 seconds while playing touch football one week ago. Since this incident, he has had intermittent vertigo, nausea, vomiting, blurred vision, a headache, and malaise. His neuro exam and CT are normal. What is the diagnosis?**

Post concussive syndrome. Most individuals recover fully over a 2 to 6 week time span. However, a few of these cases have persistent deficits.

☐ **A 16-year-old presents with progressively severe intermittent vertigo for 6 months and progressive unilateral hearing loss for 3 months. What is the diagnosis?**

Cerebellopontine angle tumor. Confirm diagnosis with a MRI scan.

☐ **Differentiate between decerebrate and decorticate posturing.**

Decerebrate posturing is when the elbows and legs are extended which is indicative of a midbrain lesion. Decorticate posturing is when the elbows are flexed and the legs are extended. This suggests a lesion in the thalamic region. Remember: DeCORticate = hands by the heart (cor).

☐ **Differentiate between focal (partial) seizures and generalized seizures.**

Partial seizures arise from a single focus and may spread out, whereas generalized seizures involve the whole cerebral cortex. Absence and grand mal seizures are examples of generalized seizures.

☐ **Recurrent seizures in patients with a history of febrile seizures generally occur in what time frame?**

About 85% occur within the first 2 years. The younger the child, the more likely recurrence will happen. If a patient has a febrile seizure in the first year of life the recurrence rate is 50%. If it occurs in the second year, the recurrence is only 25%.

☐ **What is the number one cause of seizures in a child?**

Epilepsy is the cause of 40% of seizures. The other common causes include meningitis, head trauma and stroke.

☐ **A 16-year-old girl complains of a throbbing, dull, unilateral headache that lasts for hours then goes away with sleep. She also has been nauseated and has vomited twice. She reports small areas of visual loss plus strange zigzag lines in her vision. What is the diagnosis?**

A migraine headache with an aura.

NEUROLOGY

☐ **What factors may precipitate migraine headaches?**

Bright lights, cheese, hot-dogs and other foods containing tyramine or nitrates, menstruation, monosodium glutamate, and stress.

☐ **What neoplastic process is most commonly associated with myasthenia gravis?**

Thymoma.

☐ **Which is the most common medication associated with Neuroleptic Malignant Syndrome?**

Haloperidol. Other drugs, especially antipsychotic medications, are also causative.

☐ **Describe the key signs and symptoms of a classic, a common, an ophthalmoplegic, and a hemiplegic migraine headache.**

1. Migraine without Aura: This headache is indeed the most common. It is a slow evolving headache that lasts for hours to days. A positive family history as well as two of the following are prevalent: Nausea or vomiting, throbbing quality, photophobia, unilateral pain, and increase with menses. Distinguishing feature from "Classic" migraine is the lack of visual symptoms.

2. Migraine with Aura: Prodrome lasts up to 60 minutes. Most common symptom is visual disturbance, such as homonymous hemianopsia, scintillating scotoma, fortification spectra, and photophobia. Lip, face, and hand tingling, as well as aphasia and extremity weakness may occur. Nausea and vomiting may also result.

3. Ophthalmoplegic: Currently considered a recurrent demyelinating neuropathy (recurrent painful opthalmoplegic neuropathy). Most frequently manifests in young adults. Patient has an outwardly deviated, dilated eye, with ptosis. The third, sixth, and fourth nerves are typically involved.

4. Hemiplegic: Described as a migraine with auara as well. Unilateral motor and sensory symptoms, mild hemiparesis to hemiplegia are exhibited with this type of headache. May also include vision and speech impairment.

☐ **What elements in a patients history suggest an intracranial tumor?**

(1) Wake patients from their sleep (2) worse in the morning, (3) increase in severity with postural changes or Valsalva maneuvers, (4) are associated with nausea and vomiting (5) associated with focal defects or mental status changes, and (6) occur with a new onset of seizures.

☐ **A 17-year-old woman with a history of flu like symptoms (URI) one week ago now presents with vertigo, nausea, and vomiting. No auditory impairment or focal deficits are noted. What is the most likely cause of her problem?**

Labyrinthitis or vestibular neuronitis.

PEDIATRIC EMERGENCY MEDICINE BOARD REVIEW

☐ **A 16-year-old female complains of weakness and tingling in her right arm and leg for 2 days. She reports an episode of right eye pain and blurred vision that resolved over one month, but the onset of that pain started 2 years ago. She also recalls a two week episode of intermittent blurred vision the previous year. What is the diagnosis?**

Presumptive multiple sclerosis. Confirm with MRI and CSF (look for oligoclonal bands).

☐ **What is the hallmark motor finding in Neuroleptic Malignant Syndrome?**

"Lead pipe" rigidity.

☐ **What is the most common cause of syncope in children?**

Vasovagal or simple fainting (50%).

☐ **What is the significance of bilateral nystagmus with cold caloric testing?**

It signifies that an intact cortex, midbrain, and brainstem are present.

☐ **What symptoms form the classic tetrad seen in kernicterus?**

Choreoathetosis, supernuclear ophthalmoplegia, sensorineural hearing loss, and enamel hypoplasia.

☐ **How are upper motor neuron (UMN) lesions of CN VII (facial nerve) distinguished from peripheral lesions?**

1. UMN: A unilateral weakness of the lower half of the face.
2. Peripheral: Involves the entire half of the face.

☐ **What is the most common cause of intermittent ataxia in children?**

Migraine headaches involving the basilar artery

☐ **What are the most common cause of acute ataxia (onset < 24 hours)?**

Drug toxicity and infection.

☐ **What are the three most common predisposing factors in the formation of a brain abscess?**

Cyanotic heart disease, otitis and sinusitis.

☐ **What is the most common cranial neuropathy seen in Lyme disease?**

Unilateral or bilateral facial palsy. Less frequently, the VIII nerve can also be affected.

NEUROLOGY

☐ **What are growing skull fractures?**

These are seen as a complication of skull fracture in children under three years of age. They are caused by the presence of leptomeningeal cyst that is formed by an unrecognized dural laceration. It prevents the two sides of the bone from aligning together during recovery.

☐ **When should steroids be used in the treatment of increased intracranial pressure (ICP)?**

Steroids are beneficial in the treatment of vasogenic edema, so should be used to treat increased ICP associated with tumors.

☐ **What is the main difference in the history obtained between seizures and breath-holding spells?**

Breath-holding spells are always provoked. Cyanotic breath-holding spells are provoked by crying precipitated by fright, anger, pain, or frustration. Pallid breath-holding spells are provoked by pain, especially a minor bump to the head.

☐ **Ophthalmoplegic migraine affects which cranial nerve most commonly?**

Cranial nerve III.

☐ **Which toxic neuropathy can be clinically misdiagnosed in infants as Guillain-Barré syndrome?**

Botulism. Unlike in older patients that ingest the toxin, infants are usually colonized by the bacteria. A risk factor seems to be feeding the infant honey.

☐ **What components of the history are the most important in evaluating a child with neurological problems?**

Developmental, perinatal, birth and family.

☐ **What is the significance of a mildly depressed and pulsatile fontanelle in an infant?**

Nothing. This is normal.

☐ **What is Gower's sign?**

A specific way to rise to standing from the supine position that is seen in certain forms of muscular dystrophy. The child first rolls over onto hands and knees. Then the child stands up by "walking" his/her hands up the leg.

☐ **An adolescent presents to the ED with complaints of left lower leg weakness. CSF analysis shows elevated protein but normal WBC count. What idiopathic disease might be at work here?**

Guillain-Barré syndrome.

☐ **What is the most useful test to diagnose multiple sclerosis in children?**

MRI showing multiple white matter plaques in the periventricular regions.

☐ **In a child in a coma you elicit positive doll's eyes reflex (oculocephalic reflex). Does this mean the eyes move with the head, or away from the head?**

The eyes move away from the movement of the head. This sign determines viability of the brainstem. A positive sign means the brainstem is working properly.

☐ **Which way do the eyes move in normal caloric testing?**

Remember the mnemonic COWS- Cold Opposite, Warm Same. Absence of the nystagmus occurs with eighth nerve damage.

☐ **What is the most common cause of acute onset, painless monocular blindness in patients between 15-18?**

Optic neuritis. 60% of these patients eventually develop multiple sclerosis.

☐ **How do older children with stroke present?**

Acute hemiplegia.

☐ **What are the most common causes of childhood stroke?**

Congenital heart disease and sickle cell disease.

☐ **T/F Ischemic strokes are more common in children than hemorrhagic strokes?**

False: Ischemic strokes and hemorrhagic strokes occur with equal frequency.

☐ **What are the two most common vascular etiologies of childhood stroke?**

1) Focal cerebral arteriopathy; describes unexplained focal arterial stenosis in a child with ischemic stroke. 2) Moyamoya. This is a vascular condition that leads to occlusion of intracranial arteries.

☐ **What is the most common cause of a hemorrhagic stroke in children?**

Arterial vascular malformation.

☐ **What is the only independent predictor of focal cerebral arteriopathy?**

Recent URI.

☐ **What is the incidence of stroke recurrence in sickle cell patients not receiving adequate therapy?**

50%

NEUROLOGY

☐ **What therapeutic measure decreases the recurrence of stroke in a patient with sickle cell disease?**

Chronic periodic blood transfusions and possible hydroxyurea.

☐ **A Babinski sign is indicative of an abnormality of what tract?**

Corticospinal. It is indicative of an upper motor neuron lesion.

☐ **What is the incidence of a child with a history of febrile seizures developing a convulsive disorder compared with a child who has never had a febrile seizure?**

It is slightly higher.

☐ **What is the drug of choice for simple febrile seizures?**

No treatment is usually given, aside from reducing the fever.

☐ **A child is brought in with a new onset "tic". What is your treatment?**

None at this time. Simple clonic motor tics usually resolve spontaneously in less than a year.

☐ **What can you give to ameliorate the dystonic reactions sometimes seen in a patient taking haloperidol?**

Benztropine (Cogentin), trihexyphenidyl (Artane), and diphenhydramine (Benadryl).

☐ **What metabolic abnormalities can cause neonatal seizures?**

Hypoxemia, hypoglycemia, hypocalcemia, hypomagnesemia and hyponatremia.

☐ **What is the definition of ataxia?**

Ataxia is an impairment in the coordination of movement without the loss of muscle strength.

☐ **What is the most common posterior fossa tumor?**

Astrocytoma.

☐ **Can ataxia be a manifestation of Guillain-Barré syndrome?**

Yes- if there is a significant sensory neuropathy present, there may be an impairment of afferent sensory input to the cerebellum, causing a "sensory ataxia", in addition to the muscle weakness and areflexia seen in these patients.

PEDIATRIC EMERGENCY MEDICINE BOARD REVIEW

☐ **What clinical features define a simple febrile seizure?**

1. Brief duration, usually < 15 minutes.
2. Generalized seizure.
3. Convulsions associated with the first day of an elevated temperature.
4. Occur in children between 3 mo - 5 years of age (peak incidence 12 – 18 months).
5. Does not recur within 24 hours.
6. Normal neurologic examination after the seizure.

☐ **Will a child with simple febrile seizures have a normal electroencephalogram (EEG)?**

Yes.

☐ **When should a lumbar puncture be considered in a child with a febrile seizure?**

1) Presence of meningeal signs. 2) Infants 6 months – 12 months if immunization status for S. pneumonia and H. influenza are not up to date. 3) When a patient is already on antibiotics that may mask signs and symptoms of meningitis.

☐ **What clinical features characterize children with benign paroxysmal vertigo?**

1) Age less than 4 years; 2) sudden onset of vertigo; 3) autonomic symptoms, such as nausea, vomiting and sweating; 4) nystagmus.

☐ **What symptoms and signs distinguish transverse myelitis from Guillain-Barré syndrome?**

Both transverse myelitis and acute polyneuritis share the acute onset of lower extremity weakness and paresthesias as presenting symptoms and signs. Patients with transverse myelitis will initially complain of back pain (in the absence of trauma) and urinary retention. Patients with acute polyneuritis will have bowel and bladder dysfunction later in the course. On physical examination of patients with transverse myelitis, a distinct "sensory level" will also be elicited; impairment of touch, pain and temperature sensation will be seen, but proprioception will remain intact.

☐ **What is the pathophysiology of Guillain-Barré syndrome?**

The pathologic hallmark of acute polyneuritis is demyelination of motor and sensory nerves, thought to be due to an autoimmune process, causing the classic ascending paralysis, areflexia and paresthesias.

☐ **What is the pathophysiology of myasthenia gravis?**

Antibodies against the acetylcholine receptor of the post-synaptic neuromuscular junction-this results in failure of neuromuscular transmission, with consequent fluctuating muscle weakness.

☐ **What are the presenting signs and symptoms of juvenile myasthenia gravis?**

The juvenile form of myasthenia gravis accounts for 25% of all cases and is the predominant form seen in the pediatric population, usually in school-age children. Most patients will have ptosis, oculomotor palsies and truncal or limb weakness that becomes progressively worse with continued muscle activity.

NEUROLOGY

☐ **What is the "Tensilon test"?**

The Tensilon test secures the diagnosis of myasthenia gravis. Edrophonium (Tensilon), which is an anticholinesterase drug, is given slowly by the intravenous route. Patients with myasthenia gravis will have a brief, but dramatic, resolution of their muscle weakness.

☐ **A 4-year-old presents with the acute onset of unsteadiness of gait. He has otherwise been well, except for a recent upper respiratory infection. Examination reveals ataxia, tremor and bilateral dysmetria. What is the most likely diagnosis?**

Acute cerebellar ataxia, which is characterized by the acute onset of ataxia in an otherwise healthy child, usually between the ages of 1 and 4 years. Antecedent viral infections precede the onset of symptoms in over half of patients. Resolution of symptoms usually occurs within 2 weeks of the onset of symptoms.

☐ **What must be excluded in the above patient before the diagnosis of acute cerebellar ataxia?**

A posterior fossa mass or tumor.

☐ **Is Lyme disease a common cause of seventh nerve palsy?**

Yes-there is a subset of patients with Lyme disease who present solely with an isolated seventh nerve palsy.

☐ **What is the likelihood that a patient with Bell's Palsy will completely recover?**

Complete recovery is seen in up to 80% of patients, usually within three weeks of the onset of symptoms.

☐ **What is the definition of " non-communicating" hydrocephalus?**

If the obstruction to the flow of cerebrospinal fluid occurs within the ventricular system, then non-communicating hydrocephalus results.

☐ **What is the definition of "communicating" hydrocephalus?**

If the obstruction to the flow of cerebrospinal fluid occurs outside the ventricular system at the level of any of the exit foramina or if there is an excessive production of cerebrospinal fluid, then communicating hydrocephalus results.

☐ **What is the most common cause of neonatal seizures?**

Hypoxic-ischemic encephalopathy, which accounts for nearly 60% of cases.

☐ **What agents are useful to decrease the production of cerebrospinal fluid (CSF) in patients with hydrocephalus?**

Acetazolamide, furosemide and glycerol have all been used to decrease CSF production.

☐ **What is "benign intracranial hypertension" (pseudotumor cerebri)?**

A syndrome in which patients have symptoms and signs (headache, papilledema) of increased intracranial pressure; however, physical examination shows no focal neurologic deficits or encephalopathy and there is no evidence of an intracranial mass or obstruction to the flow of CSF. CSF analysis will be normal except for increased opening pressure.

☐ **What drugs are thought to be associated with the development of benign intracranial hypertension (pseudotumor cerebri)?**

Antibiotics, such as tetracycline, minocycline, penicillin, gentamicin; oral contraceptives; growth hormones; retinoids; thyroid hormone; lithium carbonate: corticosteroid withdrawal.

☐ **What is the most common complication from a linear skull fracture?**

A subgaleal hematoma is the most common complication. These hematomas can become very large, especially if they liquefy and dissect through the subgaleal space.

☐ **A 10-year-old presents with a large scalp laceration with an underlying depressed skull fracture that needs to be elevated in the OR; however, this will not occur for several hours. The child is otherwise stable. Should the overlying laceration be repaired before the child goes to the OR?**

Yes: rarely is the site of the laceration the site of incision to repair the depressed skull fracture. Meticulous wound care and closure will prevent hemorrhage and prolonged exposure of the fracture site to skin organisms, especially if there is a delay in definitive operative treatment. Antibiotic treatment to cover skin flora is also recommended.

☐ **A child with a significant closed head injury develops generalized seizures. Which (intravenous) anticonvulsant should be administered?**

Phenytoin, as it will not cloud the sensorium or cause sedation, so that subsequent neurologic examinations can be done easily.

☐ **A 5-month-old child presents with coma. There is no history of reported trauma or viral prodrome. On examination the child is afebrile, comatose, with decerebrate posturing and fixed, dilated pupils and a bulging fontanel. What other physical examination finding would help to confirm the diagnosis?**

Funduscopic examination-if retinal hemorrhages are present, then this is most likely a case of a shaking-impact injury. A CT scan may show subdural an intracranial hemorrhage.

☐ **A child presents with severe somnolence and a two week history of gradually worsening headache. On examination he is somnolent and responsive only to pain. He also has bilateral sixth nerve palsies. What is the cause of his symptoms?**

A diffuse increase in intracranial pressure, probably the result of an intracranial mass.

NEUROLOGY

☐ **What type of breathing pattern is seen in patients with "Cheyne-Stokes" respirations?**

Patients with Cheyne-Stokes respirations will have periods of hyperpnea followed by shorter periods of apnea.

☐ **Where is the dysfunction in the CNS in patients who have Cheyne-Stokes respirations?**

Patients with Cheyne-Stokes respirations have bilateral cerebral hemisphere dysfunction with normal brainstem function.

☐ **In patients with true vertigo, at what levels of the CNS may the defect occur?**

Patients who have true vertigo have a vestibular defect that may be peripheral (involving the labyrinth in the inner ear or the vestibular nerve) or central (involving the vestibular nuclei and its central pathways).

☐ **Can otitis media be a cause of vertigo in children?**

Yes.

☐ **What is the most common cause of labyrinthitis in children?**

Viral disese like influenza, measles and mumps.

☐ **Which symptoms help to distinguish vestibular neuronitis from labyrinthitis in children?**

Children with labyrinthitis will have hearing loss, while those with vestibular neuronitis will not.

☐ **What clinical features characterize children with cyclic vomiting?**

Cyclic vomiting is a migraine variant that maybe present in young children. These patients will have repeated episodes of nausea and vomiting that can last as little as an hour but extend for days. This will always be followed by varied asymptomatic periods. This diagnosis should be considered in a school-age child with this classic on-off pattern of intense vomiting. Later in life, these children often develop migraine headaches.

☐ **What clinical features characterize children with acute confusional states?**

Acute confusional states are migraine variants that have unusual presentations, such as the onset of headache that is followed by a period of vomiting, lethargy and confusion, disorientation and unresponsiveness. These episodes of "acute confusion" may last for several hours, after which the patient may have no memory of the event. There will be a family history of migraine headache.

☐ **How does one make the diagnosis of acute confusional state?**

History of headache followed by a period of confusion or unresponsiveness that is not due to any organic cause and a family history or migraine headaches.

PEDIATRIC EMERGENCY MEDICINE BOARD REVIEW

☐ **A healthy 5-month-old infant presents with episodes of sudden flexion of the neck and upper extremities, occurring upon awakening. The child has been developing normally, the pregnancy and delivery were uneventful. The child has a normal neurologic exam, as well as a CT of the head which shows no abnormalities?**

This patient has infantile spasms, which often has an excellent prognosis for normal neuro-development.

☐ **What drug(s) is(are) preferred for the management of infantile spasms?**

Adrenocorticotrophic hormone (ACTH) and anti-seizure medicines (mainly vigabatrin).

☐ **What is the most common movement disorder in childhood?**

Transient tic disorder, which presents most often in boys. The movements consist of eye-blinking, throat clearing or other repetitive facial movements. These "tics" usually regress after one year and do not require drug treatment.

☐ **A 6-month-old infant presents with a progressively enlarging, pulsating mass on his skull. He had fallen off the changing table and hit his head on the floor one month earlier, but did not seek medical care at the time. What is the most likely diagnosis?**

A leptomeningeal cyst, which is an unusual complication of linear skull fractures. The cyst is caused by interposition of the leptomeninges and occasionally, traumatized brain, through the interrupted dura and edges of the fracture. The mass will then expand because of communication with the CSF space and the cyst.

☐ **What is Horner's syndrome?**

Miosis, ptosis, and anhydrosis (lack of sweat).

☐ **A 6-year-old child falls 5 feet, hitting his head on concrete and then has a generalized seizure lasting less than 1 minute. His neurological exam is completely normal. What is the most likely diagnosis?**

Post-traumatic seizure.

☐ **His parents want to know the likelihood of another seizure. What do you tell them?**

Approximately 25% of children with post-traumatic seizure will have additional seizures beyond one week after the injury.

☐ **A 3-year-old child presents with coma. He had intractable, non-bilious vomiting for 24 hours prior to the onset of coma. There is no history of trauma. SGOT and SGPT are elevated but the serum bilirubin is normal. What is the most likely diagnosis?**

Reye's syndrome.

☐ **What is the definition of status epilepticus?**

Continuous generalized convulsive seizures for at least 5 minutes or is repeated frequently without return to baseline consciousness between seizures.

NEUROLOGY

☐ **A 16-year-old obese girl with a history of irregular menses presents with a severe headache and blurry vision. Your physical exam is unremarkable except for papilledema. Her head CT is negative. An LP is done and opening pressure is 380 mm H2O. What is the diagnosis and treatment?**

Pseudotumor cerebri. Draining some fluid is therapeutic. Also, start her on a course of steroids.

☐ **What is the agent of choice for treating status epilepticus?**

Stopping the seizure is the primary goal, because the longer the seizure continues, the harder it is to stop. Lorazepam (Ativan) is preferred over diazepam (Valium) because of its longer duration of action.

☐ **What is the most common arrhythmia causing syncope in patients with a structurally normal heart?**

Supraventricular tachycardia (SVT).

☐ **Which MRI findings are suggestive of herpes simplex encephalitis?**

Edema or hemorrhage in the temporal lobes.

☐ **What are the features required to diagnose Guillain-Barré Syndrome?**

Progressive motor weakness and areflexia.

☐ **What is the hallmark in cerebrospinal fluid which indicate that the patient might have Guillain-Barré Syndrome?**

Albumino-cytologic dissociation, which is found as early as the third day into the illness, but may occur as late as the second or third week.

☐ **Aside from measures intended to stop seizures, what is the major therapeutic priority in the management of status epilepticus?**

Maintenance of oxygenation, adequate ventilation, and the prevention of hypoxia.

☐ **In what disease is cerebrospinal fluid albuminocytologic dissociation seen and what does it mean?**

Guillain-Barré Disease. An increase in cerebrospinal fluid protein without a corresponding increase in cerebrospinal fluid white cells is referred to as albuminocytologic dissociation.

RHEUMATOLOGY, IMMUNOLOGY AND ALLERGY

☐ **Which drug is the most common pharmaceutical cause of true allergic reactions?**

Penicillin.

☐ **Anaphylaxis-related deaths are primarily caused by penicillin. What is the second most common cause?**

Insect stings.

☐ **Most children with mucocutaneous lymph node syndrome (MLNS) are below what age?**

MLNS, or Kawasaki disease, is predominantly found in children under 5 years of age.

☐ **What percentage of patients with Kawasaki disease also develop acute carditis?**

50%, usually myocarditis with mild to moderate congestive heart failure. Pericarditis, conduction abnormalities, and valvular disturbances may occur but are less common.

☐ **A patient presents with fever, acute polyarthritis, or migratory arthritis a few weeks after a bout of Streptococcal pharyngitis. What disease should be suspected?**

Acute rheumatic fever.

☐ **What treatment should be started after the diagnosis of acute rheumatic fever has been made?**

Penicillin or erythromycin should be given even if cultures for Group A Streptococcus are negative. High-dose aspirin therapy is used at an initial dose of 75-100 mg/kg/day. Carditis or congestive heart failure is treated with prednisone, 1-2 mg/kg/day.

☐ **Myocardial infarction can occur with which two rheumatic diseases?**

Kawasaki disease and polyarteritis nodosa (PAN).

☐ **What is the most common cause of anaphylactoid reactions?**

Intravenous iodinized radiographic contrast media.

☐ **What is the treatment of choice for the patient in anaphylactic shock?**

Epinephrine (1:1000), 0.01 mg/kg IM in the anterior thigh (max dose of 0.3 mg). If hypotensive, use 0.1 mL/kg of the 1:10,000 dilution, IV.

☐ **How long should a patient with an anaphylactic reaction be observed?**

4 – 6 hours (no literature to support this). There is a worry of a biphasic reaction that can occur any time between 1 – 7 days after the initial anaphylactic episode.

☐ **What is required to make a diagnosis of Kawasaki disease, in a young patient with a prolonged fever?**

The diagnosis requires four of these five common clinical findings:
1. Conjunctival inflammation.
2. Rash.
3. Adenopathy.
4. Strawberry tongue and injection of the lips and pharynx.
5. Erythema and edema of extremities.

Desquamation of the fingers and the toes may be striking, but it is a late finding and is not one of the key clinical features of the disease.

☐ **What cardiac complication commonly occurs with SLE as well as with juvenile rheumatoid arthritis and rheumatoid arthritis?**

Pericarditis.

☐ **What is the only treatment to be proven to be of benefit in treating anaphylaxis?**

Epinephrine.

☐ **What are the articular symptoms of late, Stage III, Lyme disease?**

Chronic arthritis, especially in the knee, periostitis, and tendonitis.

☐ **A 10-year-old child presents limping and complaining of several weeks of groin, hip, and knee pain that worsens with activity. What diseases should be considered?**

Transient tenosynovitis of the hip, slipped capital femoral epiphysis, Legg-Calvé-Perthes, suppurative arthritis, rheumatic fever, juvenile idiopathic arthritis, and tuberculosis of the hip.

☐ **What disease is suspected in an adolescent with a tender, purpuric dependent rash, colicky abdominal pain, migratory polyarthritis, and microscopic hematuria?**

Henoch-Schönlein purpura, a leukoblastic vasculitis. Intestinal or pulmonary hemorrhage may occur, and 7–9% percent of the cases will develop chronic renal sequelae. Salicylates are effective for the arthritis. Other treatment is directed at the symptoms. Steroids are not particularly effective.

☐ **A child has painful swollen joints along with a spiking high fever, shaking chills, signs of pericarditis, and a pale erythematous coalescing rash on the trunk, palms, and soles. Hepatosplenomegaly is found. What is the diagnosis?**

Systemic juvenile idiopathic arthritis formally known as systemic juvenile rheumatoid arthritis, or Still's disease.

RHEUMATOLOGY, IMMUNOLOGY AND ALLERGY

☐ **What treatment besides aspirin is effective in preventing the complications of Kawasaki disease?**

Intravenous immunoglobulins can reduce the incidence of coronary artery aneurysms to less than 5%. There was a study that showed a benefit to using steroids.

☐ **What disease classically produces erythematous plaques with dusky centers and red borders resembling a bulls-eye target?**

Erythema multiforme. This disease can also produce non-pruritic urticarial lesions, petechiae, vesicles, and bullae.

☐ **Which joint is typically involved in the most common form of juvenile idiopathic arthritis?**

Knee

☐ **What autoimmune disease produces lesions that are sometimes urticarial in appearance yet are not pruritic?**

Erythema multiforme. This disease can also produce non-puritic lesions, petechiae, vesicles, and bullae.

☐ **Which drugs are most commonly implicated in toxic epidermal necrolysis?**

Sulfa drugs, NSAIDS (meloxicam), barbiturates, other antiepileptic drugs, and antibiotics.

☐ **What is the appropriate management for TEN?**

Admit for management similar to that required for extensive second-degree burns.

☐ **What can cause erythema multiforme?**

EM can be triggered by viral or bacterial infections, by drugs of nearly all classes, and by malignancy.

☐ **What is the most common cause of allergic contact dermatitis?**

Toxicodendron species, such as poison oak, poison ivy, and poison sumac are responsible for more cases of contact dermatitis than all the other allergens combined.

☐ **Why does scratching spread poison oak and poison ivy?**

The antigenic resin contaminates the hands and fingernails and is thereby spread by rubbing or scratching. A single contaminated finger can produce more than 500 reactive groups of lesions.

☐ **How is the antigen of poison oak or poison ivy inactivated?**

Careful washing with soap and water destroys the antigen. Special attention must be paid to the fingernails, otherwise the antigenic resin can be carried for weeks.

PEDIATRIC EMERGENCY MEDICINE BOARD REVIEW

☐ **What underlying illnesses should be considered in a patient with nontraumatic uveitis?**

Collagen vascular diseases, sarcoid, ankylosing spondylitis, Reiter's syndrome, tuberculosis, syphilis, toxoplasmosis, juvenile idiopathic arthritis, and Lyme disease.

☐ **What is the difference between episcleritis and scleritis?**

Both are associated with collagen vascular disorders; episcleritis is a benign superficial inflammation of the tissues between the sclera and the conjunctiva. Scleritis is a more severe and more painful inflammation of the deep sclera which can result in visual loss.

☐ **What is the probable diagnosis of a patient with myalgias, arthralgias, headache, and an annular erythematous lesion accompanied by central clearing?**

Stage I Lyme disease with the classic lesion of Erythema Chronicum Migrans (ECM).

☐ **What rheumatologic ailments can produce acute airway obstruction?**

Relapsing polychondritis due to inflammation leading to collaps, stenosis, and destruction of the tracheiobronchial tree. Also rheumatoid arthritis which is due to arthritis or edema of the cricoarytenoid joints.

☐ **What type of infections are newborns most susceptible to and why?**

Gram negative organisms. Since there is no passive transfer of IgM, which are heat stable opsonins, the opsonization process is impaired.

☐ **So then, why aren't infants more susceptible to gram positive organisms?**

Passively transferred IgG serves as effective opsonins.

☐ **What four infectious agents are most commonly associated with asplenic individuals?**

S. pneumoniae, Neisseria sp., H. influenzae and Salmonella sp.

☐ **Fatality of hereditary angioedema can occur due to what condition?**

Edema of the larynx.

☐ **What is the definition of neutropenia?**

Absolute Neutrophil Count (ANC) < 1500 cell/microliter.

☐ **Aside from chemotherapeutics which suppress bone marrow, what other agents are most commonly implicated?**

Phenothiazine, semisynthetic penicillin, NSAID's, aminopyrine derivatives, antithyroid medications.

RHEUMATOLOGY, IMMUNOLOGY AND ALLERGY

☐ **What is the most common cause of transient neutropenia?**

Viral infections which most commonly include Hepatitis A, Hepatitis B, Influenza A and B, measles, rubella, and varicella.

☐ **How long would you expect the neutropenia to persist in the setting of childhood viral infections?**

It may persist the first three to six days of the acute viral syndrome.

☐ **Neutropenia and bacterial infection may herald the onset of what clinical entity?**

Overwhelming sepsis.

☐ **What effects of corticosteroids are likely after two hours?**

Fall in peripheral eosinophils and lymphocytes.

☐ **What effects of corticosteroids are likely after six to eight hours?**

Improvement in pulmonary function in asthmatics and hyperglycemia.

☐ **Describe the clinical manifestation of urticaria?**

Well-circumscribed, erythematous raised skin lesions.

☐ **What is most common form of urticaria caused by physical factors?**

Cold urticaria.

☐ **What is the most effective treatment for control of urticaria?**

0.5 mg/kg Hydroxyzine (Atarax).

☐ **When do most anaphylactic reactions occur?**

Within the first 30 minutes after initial exposure.

☐ **What are typical initial symptoms of an anaphylactic reaction?**

A tingling sensation around the mouth, followed by a warm feeling and tightness in the chest or throat.

☐ **What is the most common manifestation of an adverse drug reaction?**

Cutaneous eruption with urticarial, exanthematous, and eczematoid types occurring predominantly.

☐ **What infectious agent is implicated in chronic blepharitis infection?**

Staphylococcal.

☐ **An 8-year-old boy is complaining of lacrimation, itching, and burning of both eyes, with significant photophobia. The patient has a past medical history significant for asthma. What is your diagnosis?**

Vernal conjunctivitis.

☐ **If you decide to treat the vernal conjunctivitis with topical steroids, what must be monitored?**

Intraocular pressure.

☐ **Define Graft Versus Host Disease (GVHD)?**

Engraftment of immunocompetent donor cells into an immunocompromised host, resulting in cell mediated cytotoxic destruction of host cells if an immunologic incompatibility exists.

☐ **When does acute GVHD present and what are the typical manifestations?**

Acute GVHD typically occurs about day 19 (median) just as the patient begins to engraft and is characterized by erythroderma, cholestatic hepatitis, and enteritis.

☐ **What are the significant categories of toxic side effects of cyclosporine therapy?**

Neurotoxic: Tremors, paraesthesia, headache, confusion, somnolence, seizures, and coma.
Hepatotoxic: Cholestasis, cholelithiasis, and hemorrhagic necrosis.
Endocrine: Ketosis, hyperprolactinemia, hypertestosteronemia, gynecomastia, and impaired spermatogenesis.
Metabolic: Hypomagnesemia, hyperuricemia, hyperglycemia, hyperkalemia, and hypocholesterolemia.
Vascular: Hypertension, vasculitic hemolytic-uremic syndrome, and atherogenesis.
Nephrotoxic: Oliguria, acute tubular damage, fluid retention, interstitial fibrosis, and tubular atrophy.

☐ **One month after renal transplant, what infectious agent most commonly causes a urinary tract infection?**

Pseudomonas aeruginosa.

☐ **A renal transplant patient thirty days post-transplant presents with fever, oliguria, hypertension, and elevated serum creatinine. What two diagnostic tests would you perform next?**

Renal ultrasound and renal scan to evaluate renal blood flow.

☐ **What are the most common manifestations of Juvenile idiopathic arthritis?**

High intermittent fever, for a minimum of two weeks duration, rheumatoid rash, arthralgia or myalgia (during febrile episodes), and persistent arthritis of greater than six weeks duration.

RHEUMATOLOGY, IMMUNOLOGY AND ALLERGY

☐ **What is the potentially irreversible sequelae of neonatal lupus?**

Congenital heart block.

☐ **List the potentially fatal manifestations of Henoch-Schönlein syndrome.**

Acute renal failure; GI complications, such as hemorrhage, intussusception, and bowel infarction; and CNS involvement, which may precipitate seizures, paresis, or coma.

☐ **What syndrome is the leading cause of acquired heart disease in the US, and what percent of cases are fatal?**

Kawasaki's disease. 1-2% or cases are fatal.

☐ **What are the major Jones criteria used to diagnose rheumatic fever?**

Carditis, chorea (Sydenham), erythema marginatum, migratory polyarthritis, and subcutaneous nodules. The diagnosis requires either 2 major or 1 major and 2 minor with evidence of previous strep infection.

☐ **What Jones criteria alone is sufficient for the diagnosis of rheumatic fever?**

Sydenham chorea. Deterioration in handwriting and increased clumsiness are commonly seen.

GENITOURINARY AND RENAL

☐ **What is the most common cause of acute renal failure?**

Acute tubular necrosis.

☐ **What is the most important feature for distinguishing between testicular torsion and epididymitis?**

The rate of the onset of pain. Torsional pain begins instantaneously at maximum intensity, whereas epididymal pain grows steadily over hours or days. Clinically, elevation of the scrotum will relieve pain related to epididymitis, but is not effective with torsional pain.

☐ **What is the most common cause of epididymitis?**

1. Prepubertal boys: Coliform bacteria.
2. In men younger than 35: Chlamydia or Neisseria gonorrhea.
3. Older than 35: Coliform bacteria.
4. Epididymitis is caused by urinary reflux, prostatitis, or urethral instrumentation.

☐ **What does epididymitis in childhood suggest?**

Obstructive or fistulous urinary defects. Epididymitis is rare in children.

☐ **What is the significance of the "blue dot" sign?**

This is pathognomonic for a torsion of the appendix testis or epididymis. With transillumination of the testis, a blue reflection occurs from the infracted tissue.

☐ **How do you manually detorse a testicle?**

Turn the testicle outward toward the thigh, like opening a book.

☐ **What percentage of patients with epididymitis will also have pyuria?**

25%.

☐ **What 4 clinical findings are indicative of acute glomerulonephritis (GN)?**

1. Oliguria.
2. Hypertension.
3. Pulmonary edema.
4. Urine sediment containing red blood cells, white blood cells, protein, and red blood cell casts.

☐ **What is the most common cause of hematuria in children presenting to the ED?**

Infection.

☐ **What are some causes of false-positive hematuria?**

Food coloring, beets, paprika, rifampin, phenothiazine, Dilantin, myoglobin, or menstruation.

☐ **A urinalysis reveals red cell casts and dysmorphic RBCs in the urine. What is the probable origin of hematuria?**

Glomerulus.

☐ **A 4-year-old boy presents with a painless mass in his scrotum that fluctuates in size with palpation. The mass transilluminates. What is the diagnosis?**

Hydrocele.

☐ **What is the composition of the most common kidney stones?**

Calcium oxalate (65%), followed by magnesium ammonium phosphate (struvite) (20%), calcium phosphate (7.5%), uric acid (5%), and crystine (1%).

☐ **What percent of urinary calculi are radiopaque?**

90%.

☐ **What are the admission criteria for patients with renal calculi?**

Infection with obstruction, a solitary kidney, uncontrolled pain, intractable emesis, or large stones (because only 10% of stones > 6 mm pass spontaneously), renal insufficiency and complete obstruction or urinary extravasation.

☐ **What percent of patients with urinary calculi will not have hematuria?**

10% - 25%.

☐ **What percent of patients spontaneously pass kidney stones?**

80%. This is largely dependent on size. 75% of stones less than 4 mm pass spontaneously, while only 10% of those larger than 6 mm pass spontaneously.

☐ **What is the most common cause of nephrotic syndrome in children?**

Minimal change disease.

GENITOURINARY AND RENAL

☐ **What is the definition of oliguria? Of anuria?**

Oliguria is as a urine output of less than 500 mL/day. Anuria is a urine output of less than 100 mL/day.

☐ **What is the initial treatment for priapism?**

Oxygen, hydration, and analgesia.

☐ **What is the most common origin of proteinuria?**

Benign orthostatic Proteinuria.

☐ **What are the risk factors for subclinical pyelonephritis?**

Multiple prior UTIs, longer duration of symptoms, recent pyelonephritis, diabetes, anatomic abnormalities, immunocompromised patients, and in indigents.

☐ **What is the most common cause of intrinsic renal failure?**

Acute tubular necrosis (80-90%), resulting from an ischemic injury.

☐ **If a urine dipstick is positive for blood yet the UA on the same urine is negative for RBCs, what is the probable disease?**

Rhabdomyolysis. Severe muscle damage can result in free myoglobin in the blood.

☐ **What is the most common neoplasm in men under 30?**

Seminomas.

☐ **Testicular torsion is most common in which age group?**

14-year-olds. Two-thirds of the cases occur in the second decade. The next most common group is newborns.

☐ **T/F: Testicular torsion frequently follows a history of strenuous physical activity or occurs during sleep.**

True.

☐ **T/F: Forty percent of patients with testicular torsion have a history of similar pain in the past that resolved spontaneously.**

True.

☐ **What is the definitive diagnostic test for testicular torsion?**

Emergent surgical exploration.

☐ **What is the definitive treatment for testicular torsion?**

Bilateral orchiopexy in which the testes are surgically attached to the scrotum.

☐ **What is the most common cause of urinary tract infections (UTI)?**

E. coli (80%). E. coli is also the most common cause of pyelonephritis and pyelitis due to its ascension from the lower urinary tract. Staphylococcus saprophyticus accounts for 5-15% of the UTI cases.

☐ **Varicoceles are most common in which part of the scrotum?**

The left. Varicoceles are a collection of veins in the scrotum.

☐ **What is the most common renal tumor in children?**

Wilms tumor. This occurs in children under 5 years of age.

☐ **What anatomic factors predispose a child to a urinary tract infection?**

Vesicoureteral reflux, obstruction, urinary stasis, and calculi.

☐ **What is the most common cause of hypertension in children?**

Chronic pyelonephritis.

☐ **What is the gold standard laboratory test for diagnosing a urinary tract infection?**

Urine culture.

☐ **A newborn presents with a palpable abdominal mass. What is the most common cause?**

A hydronephrotic kidney.

☐ **Were is the most common site of urinary tract obstruction in children?**

The ureteropelvic junction.

☐ **A neonate presents with a renal mass, what is the differential diagnosis?**

Ureteropelvic junction obstruction, multicystic renal dysplasia, solid renal tumor, and renal vein thrombosis.

☐ **What is the most common cause of a urethral stricture in a male child?**

Urethral trauma due to either iatrogenic causes (catheterization, endoscopic procedures, previous urethral reconstruction) or accidental such as straddling injuries or pelvic fractures.

GENITOURINARY AND RENAL

☐ **What is hypospadias?**

A congenital penile deformity resulting from incomplete development of the distal or anterior urethra.

☐ **What is phimosis?**

An inability to retract the prepuce at an age when it should normally be retracted.

☐ **What is a paraphimosis?**

A condition where a phimotic prepuce is retracted behind the coronal sulcus and this retraction cannot be reduced.

☐ **What are the findings in a patient with testicular torsion?**

The scrotum is swollen, tender and difficult to examine. The cremasteric reflex is absent.

☐ **What are the complications of circumcision?**

Hemorrhage, infection, dehiscence, denudation of the shaft, glandular injury, and urinary retention.

☐ **What is balanitis?**

Infection of the prepuce most often due to mixed flora.

☐ **Gross hematuria of renal origin has what characteristics?**

The color is generally brown or cola-colored and may contain red blood cells casts.

☐ **What is the most common pediatric surgery?**

Indirect inguinal hernia repair. Usually occurs on the right side in males under 1 year of age.

☐ **A 9-year-old male presents to your clinic with persistent microscopic hematuria detected on routine analysis brought on a febrile illness. Proteinuria is absent and the physical exam and laboratory evaluation are normal, what is the most likely diagnosis?**

Familial benign hematuria.

☐ **A 5-year-old child presents with sudden onset of gross hematuria, edema, hypertension, and renal insufficiency two weeks after a sore throat, what is the most likely diagnosis?**

Acute post-streptococcal glomerulonephritis.

☐ **What are the complications of acute post-streptococcal glomerulonephritis?**

Hyperkalemia, hypertension, hyperphosphatemia, hypocalcemia, acidosis, seizures, uremia, and volume overload.

☐ **What is the prognosis for a patient with post-streptococcal glomerulonephritis?**

95% of children will completely recover within one month.

☐ **What are the causes of nephrotic syndrome during the first six months of life?**

Congenital infection, (syphilis toxoplasmosis cytomegalovirus), congenital nephrotic syndrome, and diffuse mesangial sclerosis of unknown etiology.

☐ **What are the most common causes of nephrotic syndrome between six months and one year of age?**

Idiopathic nephrotic syndrome or drug-induced nephrosis.

☐ **An 11-year-old female presents with a purpuric rash on the buttocks and lower extremities, arthralgias, and abdominal pain. What is the most likely diagnosis?**

Henoch-Schönlein purpura.

☐ **A 4-year-old presents with irritability, weakness, lethargy, dehydration, edema, petechiae, and a hepatosplenomegaly ten days after an episode of gastroenteritis. Laboratory results reveal a low platelet count and hemoglobulin level, what is the most likely diagnosis?**

Hemolytic-uremic syndrome.

☐ **What are the complications of hemolytic uremic syndrome?**

Fluid overload, anemia, acidosis, hyperkalemia, congestive heart failure, hypertension, and uremia.

☐ **What is the treatment for hemolytic uremic syndrome?**

Control of fluid and electrolyte balance, control of hypertension, careful use of red cell transfusions, and early initiation of dialysis for the correction of hyperkalemia, metabolic acidosis, severe uremia, or volume overload.

☐ **A patient presents with proteinuria on routine urinalysis, which is absent, after obtaining a voiding specimen from the same patient upon awakening in the morning, what is the most likely diagnosis?**

Postural (orthostatic) proteinuria.

☐ **What is a major complication of nephrotic syndrome?**

Infection.

☐ **What is the most common type of infection in a patient with nephrotic syndrome?**

Peritonitis with strep pneumoniae.

GENITOURINARY AND RENAL

☐ **What is most common cause of interstitial nephritis in a hospitalized child?**

Medications.

☐ **What are the causes of acute renal failure in the newborn?**

Sepsis, shock, hemorrhage, renal vein thrombosis, dehydration, congenital heart disease, obstructive uropathy, renal dysgenesis, and renovascular accidents.

☐ **What are the most common causes of chronic renal failure in a child under 5 years of age?**

Anatomic abnormalities of the kidneys.

☐ **What are the most common causes of chronic renal failure in children greater than 5 years of age.**

Glomerulonephritis, hemolytic uremic syndrome, or hereditary disorders (Alport syndrome, cystic disease).

☐ **What are the causes of hypertension in a patient with chronic renal failure?**

Sodium and water overload and excessive renin production.

☐ **Which initial study should be used to differentiate obstructive from intrinsic renal causes of acute renal failure?**

Renal ultrasound.

☐ **What is the treatment for acute renal failure resulting from a pre-renal cause?**

Fluid resuscitation is the treatment of choice with an initial bolus of 20 ml/kg of crystalloid solution until vital signs become stable and urine flow is established.

☐ **What are the indications for dialysis in a patient with acute renal failure?**

BUN greater than 100 mg/dl, persistent hyperkalemia, persistent metabolic acidosis, uremic syndrome, persistent congestive heart failure, and acute renal failure with oliguria due to rhabdomyolysis.

☐ **A 6-year-old child presents with the following urine indices. Urine plasma urea nitrogen ratio greater than 8; urine: plasma creatinine ratio greater than 40; urine: plasma osmolality ratio greater than 500 mOsm/kg or greater than 1.5; and fractional excretion of sodium less than 1. What is the most likely cause of the renal failure?**

Prerenal cause such as decreased cardiac output or decreased intravascular volume due to hemorrhagic shock, dehydration, or third spacing of fluid.

☐ **A child complains of "red urine". What foods can cause this?**

Beets, blackberries and red food coloring.

☐ **What is the recommended course of action for a child with gross hematuria?**

Hospitalize and run the following tests: CBC, BUN/Cr, 24 hr creatine/protein/calcium, urine culture, serum C3, and anti-DNAse B titer. If these do not lead to a diagnosis then get an ultrasound or IVP to look for structural abnormalities.

☐ **What is the most common cause of gross hematuria in children?**

IgA nephropathy.

☐ **What is the classic presentation of post-streptococcal glomerulonephritis (PSG)?**

Sudden development of gross hematuria, hypertension, edema and renal insufficiency following a throat or skin infection with group A B-hemolytic streptococcus. Patients frequently also have generalized complaints of fever, malaise, lethargy, abdominal pain, etc.

☐ **HUS most commonly follows infection with what organism?**

E. coli (O157:H7). The disease is usually a sequelae to a bout of gastroenteritis caused by this organism. Pneumococcal pneumonia has also been associated with HUS, though this is much less common.

☐ **What is the prognosis for patients with acute renal failure secondary to HUS?**

Over 90% survival with many patients eventually recovering normal renal function.

☐ **The mother of a 2 month old male infant is concerned that she cannot retract the foreskin. Should you circumcise at this point?**

No. More than 85% of uncircumcised males can retract the foreskin by the 3rd year. Before this time it is not a concern.

☐ **What is the most common cause of acute scrotal pain in patients less than 6 years old?**

Testicular torsion.

☐ **What is the most common presentation of idiopathic nephrotic syndrome?**

Edema, frequently in conjunction with anorexia, diarrhea and abdominal pain.

☐ **Are circumcised males more, less, or equally likely to develop urinary tract infection when compared to their uncut counterparts?**

Less.

☐ **What is the most common cause of acute hemorrhagic cystitis in males?**

Adenovirus. In females, E. coli is more common, followed by adenovirus.

GENITOURINARY AND RENAL

☐ **What are the two most severe sequelae for congenital neurogenic bladder?**

Urinary incontinence and renal damage, from high intravesical pressures caused by lack of coordination of the sphincters and detrusor muscles.

☐ **What is the treatment for paraphimosis?**

As the glans edema and venous engorgement can lead to arterial compromise, this is a true emergency. First, try to manually compress the edema, after which the foreskin may be successfully reduced. If this does not work, a superficial vertical incision of the constricting band (after local anesthetic) will decompress the gland. Obviously, this procedure should be performed by a urologist or emergency medicine physician.

☐ **How long after the onset of testicular torsion do irreversible changes develop?**

About 4.5-5 hours.

☐ **During a routine examination of a 12-year-old male you encounter, in the upper scrotum, a boggy enlargement. Should you(and he) be concerned?**

No. This is most likely a varicocele, which is very common. About 15-17% of adolescent males have an asymptomatic varicocele.

☐ **How do you treat the first episode of genital herpes?**

Oral acyclovir (200 mg 5 times/day for 10 days) or valcyclovir or famciclovir.

☐ **What is the most common STD in sexually active teenage males?**

Non-gonococcal urethritis due to Chlamydia trachomatis.

☐ **Testicular US in a 15-year-old with a low-grade fever, leukocytosis, and acute onset of testicular pain shows testicular and epididymal enlargement with decreased echogenicity, scrotal skin thickening, and a hydrocele. Most likely diagnosis?**

Acute testicular torsion, also consider epididymoorchitis and tumor.

☐ **What diagnostic test should be ordered?**

Doppler US. In acute torsion, there is asymmetric or absent blood flow in the affected testicle. In epididymoorchitis, there is increased flow to the affected testicle. Flow to neoplasms is variable. Testicular torsion is the most common scrotal disorder in children. Acute epididymitis is the most common acute process in the postpubertal age group.

☐ **What age groups are at highest risk of acute testicular torsion?**

Newborn and puberty (13-16 years)

☐ **Doppler US reveals absent arterial flow in the affected testicle. How does the salvage rate depend on the onset of pain and the time of surgery?**

The salvage rate in relation to onset of pain is as follows:
80% to 100%   < 6 hours
76%   6 -12 hours
20%   12 - 24 hours
Near 0%   < 24 hours
The test is viable for 3 to 6 hours and spontaneous detorsion occurs in less than 10% of cases.

☐ **What age groups are at highest risk of acute testicular torsion?**

Newborn and puberty (13-16 years).

☐ **Which organisms are responsible for epididymoorchitis in patients younger than 35 years versus those older than 35 years?**

Escherichia coli and S. aureus. In patients younger than 35 years, Chlamydia trachomatis and Neisseria gonorrhoeae should also be considered. In older patients, E. coli and Proteus mirabilis should be suspected. Orchitis will develop in 20% of patients with mumps.

☐ **What is the leading cause of end-stage renal disease among boys?**

Posterior urethral valves. This condition is the most common cause of urinary tract obstruction among boys.

☐ **What complications may occur with posterior urethral valves?**

Neonatal urine leak, urothorax, urinoma, pneumothorax, pneumomediastinum, Prune-belly syndrome and renal dysplasia.

☐ **What is posthitis?**

Painful swelling and erythema of the prepuce (foreskin) of the penis.

HEMATOLOGY AND ONCOLOGY

☐ **Which types of blood loss are indicative of a bleeding disorder?**

1. Spontaneous bleeding from many sites.
2. Bleeding from non-traumatic sites.
3. Delayed bleeding several hours after trauma.
4. Bleeding into deep tissues or joints.

☐ **What common drugs have been implicated in predisposing to acquired bleeding disorders?**

Ethanol, ASA, NSAIDs, warfarin, and antibiotics.

☐ **Bleeding into joints or potential spaces (i.e., retroperitoneum), as well as delayed bleeding, suggests what type of bleeding disorder?**

Coagulation Factor Deficiency.

☐ **Below what platelet count is spontaneous hemorrhage likely to occur?**

< 10,000/mm3.

☐ **It is generally agreed that most patients with active bleeding and platelet counts < 50,000/mm3 should receive platelet transfusion. How much will the platelet count be raised for each unit of platelets infused?**

10,000/mm3.

☐ **What groups of patients with thrombocytopenia would be unlikely to respond to platelet infusions?**

Those with platelet antibodies (ITP or hypersplenism). Also HUS may make hemolysis worse.

☐ **What 5 treatment options are available for patients who are bleeding and have liver disease?**

1. Transfusion with PRBCS (maintains hemodynamic stability).
2. Vitamin K.
3. Fresh frozen plasma.
4. Platelet transfusion.
5. DDAVP (Desmopressin).

☐ **What are the clinical complications of DIC?**

Bleeding, thrombosis, and purpura fulminans.

☐ **What three laboratory studies would be most helpful in establishing the diagnosis of DIC?**

1. Prothrombin time—prolonged.
2. Platelet count—usually low.
3. Fibrinogen level—low.

☐ **What are the most common hemostatic abnormalities in patients infected with HIV?**

Thrombocytopenia and acquired circulating anticoagulants (causes prolongation of aPTT).

☐ **How much will one unit of Factor VIII concentrate raise the circulating Factor VIII level?**

2%.

☐ **What is the most common inherited bleeding disorder?**

Von Willebrand Disease.

☐ **70-80% of patients with von Willebrand disease have Type I. What is currently the approved mode of therapy for bleeding in these patients? What is the dose?**

DDAVP (0.3 µg/kg IV or sc q 12h for 3–4 doses).

☐ **What is the most common hemoglobin variant?**

Hemoglobin S (Valine substituted for glutamic acid in the sixth position on the ß-chain).

☐ **What types of clinical crises are seen in patients with sickle-cell disease?**

1. Vasoocclusive (thrombotic).
2. Hematologic (sequestration and aplastic).
3. Infectious.

☐ **What is the most common type of sickle-cell crisis?**

Vasoocclusive.

☐ **What percentage of patients with sickle-cell disease have gallstones?**

75% (only 10% are symptomatic).

☐ **What are the mainstays of therapy for a patient in sickle-cell crisis?**

1. Hydration.
2. Analgesia.
3. Oxygen (only beneficial if patient is hypoxic).
4. Cardiac monitoring (if patient has history of cardiac disease or is having chest pain).

HEMATOLOGY AND ONCOLOGY

☐ **What is the most commonly encountered sickle hemoglobin variant?**

Sickle-cell trait.

☐ **What is the most common human enzyme defect?**

Glucose-6-phosphate Dehydrogenase Deficiency (G-6-PD).

☐ **What is the single most useful test in ascertaining the presence of hemolysis and a normal marrow response?**

The reticulocyte count.

☐ **What is the most common morphologic abnormality of red cells in hemolytic states?**

Spherocytes.

☐ **In a child 6 months to 4 year of age with an antecedent URI, the findings of fever, acute renal failure, microangiopathic hemolytic anemia, and thrombocytopenia are suggestive of what disorder?**

Hemolytic Uremic Syndrome.

☐ **What is the most common worldwide cause of hemolytic anemia?**

Malaria.

☐ **By how much will the infusion of one unit of PRBcs raise the hemoglobin and hematocrit in a 30 kg patient?**

Hemoglobin: 1 g/dL.
Hematocrit: 3%.

☐ **What is the first step in treating all immediate transfusion reactions?**

Stop the transfusion.

☐ **What type of immediate transfusion reaction is not dose-related and can often be completed following patient evaluation and treatment with diphenhydramine?**

Allergic transfusion reaction.

☐ **What infection carries the highest risk of transmission by blood transfusion?**

Hepatitis C (1:3,300 units).

PEDIATRIC EMERGENCY MEDICINE BOARD REVIEW

☐ **What is von Willebrand's disease?**

It is an autosomal dominant disorder of platelet function. It causes bleeding from mucous membranes, menorrhagia, and increased bleeding from wounds. Patients with von Willebrand's disease have less (or dysfunctional) von Willebrand's factor.

☐ **What lab abnormalities does DIC cause?**

Increased PT, elevated fibrin split products, decreased fibrinogen and thrombocytopenia.

☐ **What factors are deficient in Classic hemophilia, Christmas disease, and von Willebrand's disease, respectively?**

Classic hemophilia: Factor VIII.
Christmas disease: Factor IX.
Von Willebrand's disease: Factor VIIIc + von Willebrand's cofactor.

☐ **Which blood product is given when the coagulation abnormality is unknown?**

FFP.

☐ **The most frequently encountered platelet disorder of childhood is:**

Idiopathic thrombocytopenic purpura.

☐ **How do gamma globulin and steroids work in the treatment of ITP?**

They block the uptake of antibody-coated platelets by splenic macrophages.

☐ **Which patients with ITP and mild head trauma without neurological findings should be treated with gamma globulin?**

Patients whose platelet count < 20,000/mm3, with signs of easy or spontaneous bleeding within one week of diagnosis, and whose follow-up is uncertain.

☐ **Typical lab findings in von Willebrand's Disease include:**

1. Normal platelet count.
2. Normal pro-time.
3. Normal or increase partial thromboplastin time.
4. Increased bleeding time.

☐ **At what age is a sickle cell patient at greatest risk for sepsis?**

Six months to 3 years.

HEMATOLOGY AND ONCOLOGY

☐ **Why is the patient from the previous question considered at high risk during this age?**

Because there is limited protective antibody, and splenic function is decreased or completely absent.

☐ **How should you treat hypoxia in a sickle cell patient who has acute chest syndrome?**

Red cell transfusion (simple or exchange).

☐ **What two processes are involved in the pathophysiology of a sickle cell patient with acute chest syndrome?**

Pulmonary infarction and/or pneumonia.

☐ **How does renal papillary necrosis that is secondary to sickle cell vaso-occlusion commonly present?**

Painless hematuria that may worsen the anemia, and red blood cells without casts in urine sediment.

☐ **Name some options for the treatment of priapism arising secondary to sickle cell vaso-occlusive crisis.**

Intravenous hydration, analgesics, red cell or exchange transfusion, and aspiration of the penile corpora.

☐ **What are some of the signs and symptoms of central nervous system infarction secondary to sickle cell vaso-occlusion?**

Mild, fleeting TIA-like symptoms, seizures, hemiparesis, coma, death.

☐ **What are the signs and symptoms of splenic sequestration crisis?**

Pallor, weakness, lethargy, disorientation, shock, decreased level of consciousness, and an enlarged spleen.

☐ **How do you treat splenic sequestration crisis?**

Rapid infusion of saline, and transfusion of red cells or whole blood.

☐ **Which factors contribute to physiologic anemia during infancy?**

Abrupt cessation of erythropoiesis with onset of respiration, low erythropoietin levels (made in the liver during the neonatal period), decreased half life and increased volume of distribution of erythropoietin, shortened survival of fetal RBCs, and expansion of the infant's blood volume with rapid weight gain during the first three months.

☐ **What are the most common sites for bleeding in patients who are hemophiliacs?**

Joints, muscles, subcutaneous tissue.

PEDIATRIC EMERGENCY MEDICINE BOARD REVIEW

☐ **What serious sequela of hip hemarthrosis is common among hemophiliacs?**

Aseptic necrosis of the femoral head.

☐ **A hemophiliac presents with pain in the abdomen and groin, flexion of the thigh, and paresthesia of the lower thigh. What condition do you suspect?**

Iliopsoas hemorrhage.

☐ **A hemophiliac presents with a headache after minor head trauma. What should your management include?**

CT of the brain, and factor replacement.

☐ **How may an oncologic patient with a serum calcium concentration of 13 mgs/dl may be treated?**

Furosemide and forced saline diuresis. Extreme hypercalcemia may require mithramycin.

☐ **The most common form of childhood malignancy is?**

Leukemia.

☐ **What is the best treatment for superior vena cava syndrome?**

Emergent radiation therapy to shrink the tumor.

☐ **The treatment for a neutropenic febrile leukemic patient includes what?**

A semisynthetic antipseudomonal penicillin, an aminoglycoside, plus a first generation cephalosporin for gram positive coverage. Ceftazidime may be used by itself, as can cefepime.

☐ **A 2-year-old male is brought in by his mother because he has a flank mass. He is hypertensive and polycythemic. Urinalysis reveals hematuria. What is your diagnosis?**

Wilms tumor.

☐ **A 15-year-old male complains of tibial pain after playing football. X-ray shows an "onion skin" appearance of the tibia. What condition is the patient likely to have?**

Ewing Sarcoma. The onion skin appearance is characteristic of this tumor.

☐ **What is the leading cause of death from leukemia?**

Infection. The second leading cause is hemorrhage.

HEMATOLOGY AND ONCOLOGY

☐ **What are the most common malignancies to cause tumor lysis syndrome?**

Advanced leukemia and non-Hodgkin's lymphoma. Large tumor burdens with aggressive chemotherapy are risk factors.

☐ **Acute renal failure, Coomb's negative hemolytic anemia, and thrombocytopenia comprise the triad for which syndrome?**

Hemolytic-Uremic Syndrome (HUS).

☐ **Aplastic crisis in sickle cell disease is associated with what condition?**

Parvo virus. Primary erythropoietic failure can result in life-threatening anemia in a patient with a significantly decreased RBC life span.

☐ **Primary or spontaneous bacterial peritonitis is more frequently seen with patients who have which underlying conditions?**

Nephrotic syndrome, chronic liver disease, and systemic lupus erythematosus.

☐ **What symptoms characterize tumor lysis syndrome?**

Hyperuricemia, hyperkalemia, and hyperphosphatemia.

☐ **The major goal of therapy for tumor lysis syndrome is to prevent?**

Acute renal failure. The principles for treatment include alkalinization of the urine, vigorous hydration and diuresis, and prevention of the formation of toxic metabolites.

☐ **What is the treatment of choice for a brain tumor edema?**

Dexamethasone.

☐ **A four-year-old sickle cell patient presents with pallor, weakness, tachycardia and abdominal fullness. What should you suspect?**

Acute splenic sequestration.

☐ **How should you treat the patient characterized in the previous question?**

Colloid and whole blood transfusion for correction of hypovolemia, anemia and prevention of circulatory failure.

☐ **How many subsequent episodes of acute splenic sequestration be avoided?**

Splenectomy or chronic transfusions.

DERMATOLOGY

☐ **Are the nodules of erythema nodosum most often symmetrically or asymmetrically distributed?**

Erythema nodosum produces distinctive bilateral tender nodules with underlying red or purple shiny patches of skin that develop most commonly in a symmetric distribution along the extensor surfaces of the lower legs, arms, thighs, calves, and buttocks. However, the lesions may appear on any surface.

☐ **What therapy is effective for treating erythema nodosum?**

There is no therapy known to alter the course of the disease. It is a self-limited disease that may last several weeks and requires only symptomatic relief using NSAIDs, cool wet compresses, elevation, and bed rest. For severe cases prednisone may be considered.

☐ **Where are Candida albicans infections of the skin are most commonly located?**

In the intertriginous areas (i.e., in the folds of the skin, axilla, groin, under the breasts, etc.) Candida albicans appears as a beefy red rash with satellite lesions that may coalesce into larger lesions.

☐ **What is a carbuncle?**

A deep abscess that interconnects and extends into the subcutaneous tissue. Commonly seen in patients with diabetes, folliculitis, steroid use, obesity, heavy perspiration, and in areas of friction.

☐ **A mother is worried that her 5-year-old will get chicken pox because she was playing with her neighbor who was recently diagnosed with the disease. She was not immunized. At the time, the neighbor was noted to have crusty lesions all over his body. Would she be expected to develop chicken pox herself if this was the only day she played with the neighbor?**

No. Chicken pox is only contagious 48 hours before the rash breaks out and until the vesicles have crusted over.

☐ **A mother brings her 14-year-old boy to you a week after you prescribed ampicillin for his pharyngitis. Mom says he quickly began to developed a rash over his torso, arms, legs, and even the palms of his hands. On examination the patient has a erythematous, maculopapular rash in the areas described. What other diagnosis should be considered other than pharyngitis?**

Infectious mononucleosis.

☐ **Ecthyma most commonly presents on what body parts?**

The lower extremities. Ecthyma is similar to impetigo but can also be associated with a fever and lymphadenopathy. It is caused by group A beta-hemolytic streptococci.

☐ **What is the most common cause of erythema multiforme?**

Herpes simplex infections (90%).

☐ **What is the most common location of erythema nodosum?**

The shins. They can also be found on the extensor surfaces of the forearms.

☐ **A patient presents with a raised, red, small, and painful plaque on the face. On exam, a distinct, sharp, advancing edge is noted. What is the disorder and cause?**

Erysipelas. It is caused by group A streptococci.

☐ **What are the causes of exfoliative dermatitis?**

Chemicals, drugs, and cutaneous or systemic diseases.

☐ **What is a furuncle?**

Skin abscesses caused by staphylococcal infection, which involve a hair follicle and surrounding tissue.

☐ **What is hidradenitis suppurativa?**

Chronic suppurative abscesses that occur in the apocrine sweat glands of the groin and/or axilla.

☐ **Your neighbor brings her 4-year-old little girl to you because she has a "disgusting" rash. The child's face is patched with vesiculopustular lesions covered in a thick, honey-colored crust. Just two days ago, these lesions were small red papules. What is the most likely diagnosis?**

Impetigo contagiosa.

☐ **What is the most common organism responsible for the above child's infection?**

Staphylococcus aureus. Beta-Hemolytic Streptococcus is the second most common infecting agent.

☐ **What is the Koebner phenomenon?**

The development of plaques in areas where trauma has occurred. This is most common in patients with psoriasis and lichen planus.

☐ **What is a pilonidal abscess?**

An abscess which occurs in the gluteal fold as a result of disruption of the epithelial surface.

☐ **Where does a peri-rectal abscess originate?**

Anal crypts and burrows through the ischiorectal space.

DERMATOLOGY

☐ **A 12-year-old female comes to your office complaining of intense itching in the webs between her fingers that worsens at night. On close examination you notice a 1 cm long raised see a few small squiggly lines 1 cm x 1 mm where the patient has been scratching herself. Diagnosis?**

Scabies. Caused by the mite Sarcoptes scabiei var. hominis. Scabies are spread by close contact; therefore, all household contacts should also be treated.

☐ **Where is the most common site of eruption of herpes zoster?**

The thorax.

☐ **A 14-year-old female comes to your office complaining of a painful red rash. On exam, you note crops of blisters on erythematous bases in a band-like distribution across on the right side of her lower back, spreading down and out towards her hip. What is the most likely Diagnosis?**

Herpes zoster disease also known as shingles.

☐ **A patient with shingles extending to the tip of his nose is at risk for what?**

Herpetic ophthalmia resulting in corneal ulceration and scarring. Lesions on the tip of the nose indicate that the nasociliary branch of the ophthalmic nerve is affected and the cornea is at risk. This is a medical emergency and needs immediate referral to an ophthalmologist.

☐ **What are the 2 distinct causes of toxic epidermal necrolysis (scalded skin syndrome)?**

TEN is most commonly drug induced however it may also be caused by Staphylococcal infection.

☐ **What body areas are most commonly affected by staphylococcal scalded skin syndrome (SSSS)?**

The nose, mouth, neck, axillae, and groin.

☐ **What is the treatment for SSSS?**

Oral or IV penicillinase-resistant penicillin, fluid resuscitation and wound care similar to thermal burns including baths of potassium permanganate or dressings soaked in 0.5% silver nitrate, and fluids.

☐ **What is the most common cause of Stevens-Johnson syndrome?**

Drugs, most commonly sulfa drugs.

☐ **Tinea capitis is most commonly seen in what age group?**

Children aged 4–14. This is a fungal infection of the scalp that begins as a papule around one hair shaft and then spreads to other follicles. Trichophyton tonsurans is responsible for 90% of the cases.

☐ **What are the two most common causes of bullous impetigo?**

Staphyloccus aureus and streptpcoccus pyogenes.

☐ **What is the best treatment for impetigo?**

Mupirocin (Bactroban) 2% ointment and cephalexin, clindamycin, dicloxacillin, or amoxicillin/clavulanate.

☐ **What is the treatment for atopic dermatitis?**

Oral antihistamines, topical steroids, emollients, and avoidance of irritants.

☐ **What is the recommended therapy for tinea capitis?**

Oral Griseofulvin (20-25 mg/kg/day) for 8 weeks. Selenium or ketoconazole shampoos are adjunct treatments.

☐ **What would you call a tender, fluctuant mass located on the head of a child with known tinea capitis?**

A kerion.

☐ **What is the causative agent of tinea versicolor?**

Malassezia globosa fungus and in a small number of cases malassezia furfur.

☐ **What is the drug of choice for head lice?**

Permethrin (Nix), 1% lotion shampoo.

☐ **What is the treatment of choice for scabies?**

5% Permethrin cream. for the patient and family members.

☐ **What is the life expectancy of a newborn who develops erythema toxicum three days after his birth?**

Normal. This is a benign and self-limited disease that is extremely common in the neonate.

☐ **An 11-year-old female presents with a maculopapular rash on trunk and upper legs. The rash is in lines along the long axis of the ovoid lesions. The patient states that 1 week ago, she noted a single lesion which was round and 1 inch wide. What is your diagnosis?**

Pityriasis rosea.

☐ **What is the distribution of the rash in pityriasis rosea?**

"Christmas tree" pattern with parallel rows of lesions on the trunk that follow Langer lines.

☐ **Is there a role for topical steroids in acute severe sunburn?**

Yes. It may relieve pain and inflammation.

DERMATOLOGY

☐ **Patients on which antibiotic are most likely to have photoallergic drug reaction?**

Tetracycline and sulfonamides.

☐ **A 17-year-old girl complains that she can't wear her gold earring without developing a localized rash. The same rash tends to occur around all of her pierced body parts (nose, navel, and eyebrows). What is the patient most likely allergic to?**

Nickel, which is commonly found in some jewelry. She should be advised to wear either stainless steel or pure gold to avoid the allergy.

☐ **In a child with scalded skin syndrome (SSS), you notice that pressing on the blister enlarges it laterally. What is the name of this phenomenon?**

Nikolsky sign which idicates epidermal fragility.

☐ **What is the treatment for scalp lesions of seborrheic dermatitis?**

Salicylic acid shampoo, or application of mineral oil, followed by washing and removal of scales with a comb.

☐ **Can you develop Rhus dermatitis (aka allergic dermatitis following exposure to poison ivy, poison oak, and/or poison sumac) from physical contact with someone who already has the condition?**

Not unless they have not yet washed off the resin which contains the allergen.

SURGICAL EMERGENCIES

☐ **How many days should sutures remain in the following areas: face, scalp, trunk, hands and back, and the extremities?**

Neck: 3-4; Face: 5; scalp: 5-7; trunk/upper extremities: 7-10; lower extremities: 8-10; joint surface: 10-14.

☐ **How should hair be removed prior to wound repair?**

Clip the hair around the wound. Do not use a razor during the preparation, as it will increase the infection rate.

☐ **Why is epinephrine added to local anesthesia?**

To increase the duration of the anesthesia and reduce systemic absorption. However, epinephrine also causes vasoconstriction and decreased bleeding, which weakens tissue defenses and increases the incidence of wound infection.

☐ **Lidocaine with epinephrine is a useful combination for anesthesia and the cessation of bleeding. On which parts of the body is this mixture contraindicated?**

The nose, ears, digits, and penis. Epinephrine causes local constriction of the blood vessels, which can also increase infection.

☐ **What are the maximum doses of lidocaine with and without epinephrine?**

Maximum dose of lidocaine without epinephrine is 3 mg/kg
Maximum dose of lidocaine with epinephrine is 7mg/kg

☐ **What is the rate of wound infection in the average surgical service?**

4–7%. Clean wound infection rate is 2%, clean-contaminated wounds 4-10%, contaminated wounds 10–15%, and dirty wounds 25–40%. Wound infection is dependent on the degree of contamination, viability of tissue, blood supply to the tissue, dead space, amount of foreign material, age, concomitant infections, nutrition, and immunocompromised status.

☐ **What mechanisms of injury create wounds that are most susceptible to infection?**

Compression or tension injuries. They are 100 times more susceptible to infection.

☐ **A 15-year-old stepped on a nail that went right through her shoe and into her foot. Aside from normal skin flora, what organism might infect her puncture wound?**

Pseudomonas aeruginosa.

☐ **Which has a greater resistance to infection, sutures or staples?**

Staples. Braided sutures are especially prone to infection.

☐ **What is the amount of bacteria necessary to cause wound infection, with and without a foreign body?**

With foreign body: 100 bacteria.
Without foreign body: > 1 million bacteria.

☐ **Which two factors determine the ultimate appearance of a scar?**

Static and dynamic skin tension on surrounding skin.

☐ **What are the characteristics of tetanus prone wounds?**

Age of wound:	> 6 hours
Configuration:	Stellate wound
Depth:	> 1 cm
Mechanism of injury:	Missile, crush, burn, frostbites
Signs of infection:	Present
Devitalized tissue:	Present
Contaminants:	Present
Denervated and/or ischemic tissue:	Present

☐ **What is the length of time in which a wound can be closed primarily?**

The "golden period" is within 6 hours, however a wound with low infection risk can be closed within 12-24 hours.

☐ **How long should one wait before delayed primary closure (tertiary closure)?**

3-5 days. This will decrease the infection rate; it is used for severely contaminated wounds.

☐ **How long should an area of abraded skin or a laceration be kept out of the sun?**

6 months. Abraded skin will develop permanent hyperpigmentation after exposure to the sun.

☐ **What is the most common organism in wound infections?**

Staphylococcus.

☐ **How much fluid needs to collect in the chest if it is to be seen on a decubitus or upright chest x-ray?]**

200–300 ml; if supine, greater than 1 L may be necessary to be seen on AP CXR.

SURGICAL EMERGENCIES

☐ **What is Kehr's sign?**

A pain in the left shoulder made worse in Trendelenburg. This is a sign of splenic injury or inflammation.

☐ **Which organisms are most commonly responsible for overwhelming postsplenectomy sepsis?**

Encapsulated organisms: Pneumococcus (50%); Meningococcus (12%); E. Coli (11%); H. Influenzae (8%); Staphylococcus (8%); and Streptococcus (7%).

☐ **What is the most common acute surgical condition of the abdomen?**

Acute appendicitis.

☐ **Why would you order a chest x-ray in a case of acute abdomen?**

An upright chest x-ray can more easily show air under the diaphragm, an indication of a ruptured viscous.

☐ **What might the abdominal films reveal on a patient with appendicitis?**

Appendocolith (8-10%, pathognomonic), localized ileus with air fluid levels, sentinel loops with air fluid levels in the RLQ, more commonly a non-specific bowel gas pattern.

☐ **What is the most common cause of appendicitis?**

Fecaliths. Fecaliths are found in 40% of uncomplicated appendicitis cases, 65% of cases involving gangrenous appendices that have not ruptured, and in 90% of cases involving ruptured appendices.

☐ **How does retrocecal appendicitis most commonly present?**

Dysuria and hematuria (due to the proximity of the appendix to the right ureter). Poorly localized abdominal pain, anorexia, nausea, vomiting, diarrhea, mild fever, and peritoneal signs.

☐ **Differentiate between McBurney's point, Rovsing's sign, the obturator sign, and the psoas sign.**

McBurney's point	Point of maximal tenderness in a patient with appendicitis. Location is 2/3 of the way between the umbilicus and the iliac crest on the right side of the abdomen.
Rovsing's sign	Palpation of LLQ causes pain in the RLQ.
Obturator sign	Internal rotation of a flexed hip causes pain.
Psoas sign	Extension of the right thigh causes pain. This is also indicative of an inflamed appendix.

☐ **What is the higher perforation rate in appendicitis?**

The pediatric population has a perforation rate of 15–50% depending on age. Children <5yr 51-100% due to thinner omentum. Associated mortality rate of 3%.

PEDIATRIC EMERGENCY MEDICINE BOARD REVIEW

☐ **What are Grey-Turner's and Cullen's signs?**

Grey-Turner's sign–flank ecchymosis indicative of pancreatic disease.
Cullen's sign- periumbilical ecchymosis indicative of pancreatic disease. Both are caused by dissection of blood retroperitoneally.

☐ **What is the most common cause of pancreatic pseudocysts in children?**

Trauma. Pseudocysts are generally filled with pancreatic enzymes and are sterile. They appear as upper abdominal pain with anorexia, emesis, and weight loss. Evolve over several weeks after a bout of acute pancreatitis. 40% of them regress on their own.

☐ **What is the difference between cholelithiasis, cholecystitis, choledocholithiasis, and cholangitis?**

Cholelithiasis:	The existence of gallstones in the gallbladder.
Cholecystitis:	The inflammation of the gallbladder secondary to gallstones.
Choledocholithiasis:	The existence of gallstones in the common bile duct that have migrated from the gallbladder.
Cholangitis	The inflammation of the common bile duct often secondary to bacterial infection or choledocholithiasis.

☐ **Where do most hernias occur?**

In the groin (75%). Incisional and ventral hernias account for 10%, and umbilical hernias account for 3%.

☐ **The majority of inguinal hernias in infants and children are what kind of hernias?**

Indirect inguinal hernias.

☐ **Differentiate between reducible, incarcerated, and strangulated.**

Reducible: The contents of the hernia sac return to the abdomen spontaneously, or with slight pressure, when the patient is in a recumbent position.
Incarcerated: The contents of the hernia sac are irreducible and cannot be returned to the abdomen.
Strangulated: The sac has compromised blood flow. Can occur with 24hr without reduction.

☐ **What is the most common hernia in females?**

Direct hernias.

☐ **Of all the hernias in the groin area, which type will most likely strangulate?**

A femoral hernia.

☐ **How do you differentiate between an inguinal hernia and a hydrocele?**

Hydrocele will transilluminate. No mass will be felt in area of internal ring and the examiner will be able to get their fingers around the hydrocele since it does not communicate with abdominal cavity.

SURGICAL EMERGENCIES

☐ **What is the method of reducing an inguinal herniea?**

Taxis. Sedate child, exert mild pressure at internal ring and gentle traction to elongate sac, guide back into abdominal cavity.

☐ **What differentiates Hirschsprung's disease from functional constipation?**

Hirschsprung's is present from birth, no abdominal pain, encoparesis is absent, abdominal distension can be impressive, rectal exam with empty vault, abnormal motility, no ganglion cells on biopsy.

☐ **What is the major cause of death in patients with Hirschsprung's disease?**

Enterocolitis.

☐ **What is the treatment for a subungual hematoma?**

Nail trephination with opening >3mm to allow ongoing drainage without risk of clotting, nail removal with nail bed repair if nail or margins disrupted or if there is displaced phalangeal fracture.

☐ **What is the treatment for an ingrown toenail?**

Wedge resection in which the ingrown part of the nail is cut at a 30° angle. This treatment is recommended only after a conservative approach of soaking in warm water and hydrogen peroxide, wide shoes, good hygiene, and a course of antibiotics has been tried.

☐ **What is a felon and what is the treatment for this disorder?**

Deep infection of distal pulp space of fingertip, usually Staphylococcus aureus. Treat by incising the pulp space.

TRAUMA

☐ **A patient opens his eyes to voice, makes incomprehensible sounds, and withdraws to painful stimulus. What is his GCS?**

9.

Glasgow coma scale:

EYE OPENING	BEST VERBAL RESPONSE	BEST MOTOR RESPONSE
4 spontaneously	5 oriented x 3	5 obeys command
3 on request	4 confused conversation	4 localizes pain stimulus
2 to pain	3 words spoken	3 flexes either arm
1 no opening	2 groans to pain	2 extends arm to pain
	1 no response	1 no response

☐ **Can significant head trauma, causing intracranial bleeding, lead to hypovolemic shock in a child?**

As a general rule, closed head trauma, no matter how significant, rarely causes hypovolemic shock in children. However, children less than 12 months of age who have large epidural hematomas with overlying skull fractures may decompress the hematoma into the scalp to a degree that significant blood loss can occur. This is the one rare exception to the rule.

☐ **A 15-year-old fire victim is burned over both legs, his entire back, and his right arm. What percentage of his body is burned?**

63%. Follow the adult rule of 9s. Face = 9%. Arms = 9% each. Front = 18%. Back = 18%. Legs = 18% each.

☐ **A 14-year-old patient who has been burned over the entire top of his body (arms and torso, front and back) develops severe difficulty breathing and appears to be going into respiratory arrest. What should you do?**

Perform an escharotomy. The patient is most likely suffering ventilatory restriction due to the circumferential eschar about his chest that is constricting the chest cavity. Escharotomies need not be performed with anesthesia, not even local anesthesia, because third degree burns involve the nervous tissue and are thus insensitive to pain.

☐ **What is the initial treatment of tension pneumothorax?**

Large bore IV catheter placed in the second intercostal space (not a chest tube).

☐ **How much fluid needs to collect in the chest if it is to be seen on a decubitus or upright chest x-ray?**

200–300 ml; if supine, greater than 1 L may be necessary to be seen on AP CXR.

☐ **What should be checked prior to inserting a chest tube in an intubated patient with respiratory distress and decreased breath sounds on one side?**

Position of the ET tube.

☐ **When does a subdural hematoma become isodense?**

Between 1–3 weeks after the bleed. At that time, it may not show up on a CT unless you use contrast.

☐ **What percentage of C-spine fractures will be identified on a lateral x-ray?**

90%.

☐ **Describe the key features of spinal shock.**

Sudden areflexia which is transient and distal and lasts hours to weeks. BP is usually 80–100 mmHg with paradoxical bradycardia.

☐ **The corneal reflex tests:**

Ophthalmic branch (V1) of the trigeminal (fifth) nerve afferent and the facial (seventh) nerve efferent.

☐ **Define increased intracranial pressure.**

ICP > 15 mmHg.

☐ **What are the classes of shock?**

	CLASS I	CLASS II	CLASS III	CLASS IV
BLOOD VOLUME LOSS	<15%	15%-30%	30%-40%	>40%
PULSE RATE	NORMAL	MILD TACHYCARDIA	MODERATE TACHYCARDIA	SEVERE TACHYCADIA
BLOOD PRESSURE	NORMAL/ INCREASED	NORMAL/ DECREASED	DECREASED	DECREASED
RESPIRATORY RATE	NORMAL	MILD TACHYPNEA	MODERATE TACHYPNEA	SEVERE TACHYPNEA
URINE OUTPUT	1-2 ML/KG/H	0.5-1.0 ML/KG/H	0.25-0.5 ML/KG/H	NEGLIGIBLE
MENTAL STATUS	SLIGHTLY ANXIOUS	ANXIOUS	ANXIOUS/ CONFUSED	CONFUSED/ LETHARGIC

TRAUMA

☐ **What is the most sensitive indicator of shock in children?**

Tachycardia.

☐ **What is the initial fluid bolus that should be given to children in shock?**

20 ml/kg.

☐ **Do post-traumatic seizures occur more frequently in children or young adults?**

Children.

☐ **What are the most common lower extremity injuries to bone in children?**

Tibial and fibular shaft fractures, usually secondary to twist forces.

☐ **What is the basic disorder contributing to the pathophysiology of compartment syndrome?**

Increased pressure within closed tissue spaces compromising blood flow to muscle and nerve tissue. There are three prerequisites to the development of compartment syndrome:
1. Limiting space.
2. Increased tissue pressure.
3. Decreased tissue perfusion.

☐ **What are the two basic mechanisms for elevated compartment pressure?**

1. External compression—by burn eschar, circumferential casts, dressings, or pneumatic pressure garments.
2. Volume increase within the compartment—hemorrhage into the compartment, IV infiltration, or edema due to post-ischemic (postischial) swelling or secondary to injury.

☐ **What are the early signs and symptoms of compartment syndrome?**

1. Tenderness and pain out of proportion to the injury.
2. Pain with active and passive motion.
3. Hypesthesia (paresthesia)—abnormal two point discrimination.

☐ **What are the late signs and symptoms of compartment syndrome?**

1. Compartment is tense, indurated, and erythematous.
2. Slow capillary refill.
3. Pallor and pulselessness.

☐ **Where is the most common site of compartment syndrome?**

Anterior compartment of the leg.

☐ **Which two fractures are most commonly associated with compartment syndrome?**

Tibia (resulting most often in anterior compartment involvement), and supracondylar humerus fractures.

☐ **What is Volkmann's ischemia?**

It is pain worsened by passive motion of a muscle and associated neurologic deficits secondary to ischemia of the muscles and nerves; it is usually associated with a compartment syndrome.

☐ **What is Volkmann's contracture?**

Contracture deformity of fingers and sometimes wrist secondary to muscle necrosis after improper use of tourniquet or severe injury to elbow. A similar condition can develop in lower leg and foot after vascular damage to leg muscles.

☐ **What is the risk associated with not treating a septal hematoma of the nose?**

Aseptic necrosis followed by absorption of the septal cartilage, resulting in septal perforation.

☐ **What are the important structures to be considered when repairing lacerations of the cheek?**

Facial nerve, parotid duct, and gland.

☐ **What procedure should be considered as an alternative to cricothyrotomy in children?**

Needle cricothyrotomy with jet insufflation. Cricothyrotomy should is not recommended in children <10yr.

☐ **What are the features of central cord syndrome?**

Loss of motor function, worse in the upper extremities than the lower extremities. There is often sparing of perianal sensation. It is usually caused by hyperextension of the neck. ~2/3 will completely recover.

☐ **What are the features of the Brown-Sequard syndrome?**

Ipsilateral motor deficit and contralateral loss of pain and temperature sensation. Caused by penetrating injury to one side of the spinal cord. Light touch is usually absent on the side of the lesion. ~1/3 will recover.

☐ **Which pattern of partial cord injury has the worst rate of functional recovery? What are it's features?**

Anterior cord syndrome. Motor function lost below level of lesion, touch and proprioception preserved. Caused by damage to anterior spinal artery.

☐ **At what level of the cervical vertebrae is pseudosubluxation most commonly seen in children?**

C2 on C3, and C3 on C4.

TRAUMA

☐ **What is the upper limit for the normal Atlas-dens interval on a pediatric C-spine x-ray?**

4 mm. In an adult the space should be no wider than 2-3 mm.

☐ **How is Cerebral Perfusion Pressure calculated?**

CPP=MAP-ICP. Ideal CPP is 50-60 mm Hg.

☐ **What is the most likely cause of a subdural hematoma with retinal hemorrhages in a child?**

Child abuse.

☐ **A child is intubated following a closed head injury. Hyperventilation should be instituted with what target arterial PCO2?**

30-35 mm Hg.

☐ **What percent of pediatric burns are infected?**

16-20%.

☐ **A child sustains an electrical burn to the corner of his mouth from chewing an electrical cord. What complication should be anticipated?**

Delayed hemorrhage from the labial artery in 1-2 weeks.

☐ **What clinical features differentiate partial thickness from full thickness burns?**

Partial thickness: Pink to mottled red, blisters, bullae or moist weeping surface, painful.
Full thickness: Waxy white, charred or dark red, dry leathery, insensate.

☐ **In estimating the size of a burn, which body part varies most as a percent body surface area with age?**

The head. At birth the head accounts for 19% BSA. At age 1-4 it is 17%, age 5-9=13%, age 10-14=11%, age 15=9%, adult =7%.

☐ **Which fractures are thought to be specific for child abuse?**

Metaphyseal fractures (chip fractures in a corner), posterior rib fractures, acromioclavicular fractures, and multiple fractures of different ages.

☐ **Which bony lesions suggest possible child abuse?**

Spinal fractures, digital fractures, complex skull fractures, spiral long bone fractures, scapular fractures, sternal fractures, metaphyseal fractures, and periosteal separation.

PEDIATRIC EMERGENCY MEDICINE BOARD REVIEW

☐ **What findings comprise the shaken baby syndrome?**

Retinal hemorrhage, subdural hematoma, and diffuse brain swelling.

☐ **What is the leading cause of death among children aged 1 - 14 years in the United States?**

Motor vehicle traffic injuries.

☐ **What are the three most common failures in pediatric trauma resuscitation?**

Failure to open and maintain an airway.
Failure to provide appropriate fluid resuscitation.
Failure to recognize and treat internal hemorrhage.

☐ **At what point should a qualified surgeon be involved in pediatric trauma care?**

As early as possible in the course of resuscitation.

☐ **How can significant mortality risk be defined?**

Pediatric Trauma Score of 8 or less or Revised Trauma Score of 11 or less.

☐ **What special consideration must be taken in opening the airway of a multisystem trauma victim?**

Control of the cervical spine.

☐ **T/F: Cervical spine injury is less common in pediatric than adult trauma.**

True, because the child's spine is more elastic and mobile than that of the adult, and the softer pediatric vertebrae are less likely to fracture with minor stress.

☐ **Spinal cord damage secondary to acceleration-deceleration injury is usually secondary to what spinal injury?**

Subluxation, most often at the atlantooccipital base (base of skull-C1) or atlantoaxial (C1-C2) joints in infants and toddlers or the lower (C5-C7) cervical spine in school-age children.

☐ **What is SCIWORA?**

Spinal Cord Injury Without Radiographic Abnormality. This can be seen at any age, but more commonly in children under 8.

☐ **What is it about the pediatric spine that permits SCIWORA?**

Its increased elasticity and mobility caused by relative laxity of the cervical spine ligaments, incomplete development of the cervical musculature, and the shallow orientation of facet joints in young children.

TRAUMA

☐ **Because of the recognition of SCIWORA as an important cause of pediatric spinal cord injury, what test used in adults to rule out injury in the cervical spine in adult blunt trauma victims cannot rule out such injury in children?**

The lateral cervical spine x-ray.

☐ **As a result of the recognition of SCIWORA, what special precautions must be taken in all children with multiple injuries, especially those who are apneic?**

Precautions to avoid potential exacerbation of cervical spine injury in each phase of airway management and control.

☐ **T/F: Traction must be maintained on the neck at all times.**

False. Neutral stabilization, never traction.

☐ **At what point may an oral airway be used?**

After the airway has been effectively opened and the cervical spine has been simultaneously stabilized, an oral airway may be placed in an unconscious patient.

☐ **Why is correct semirigid extrication collar sizing important?**

Because excessively large collars may allow neck flexion or hyperextension.

☐ **Why should soft collars be avoided?**

They do not immobilize the cervical spine effectively and have no role in initial stabilization of the child with potential cervical spine injuries.

☐ **Why is the pediatric airway difficult to control?**

It is narrow and easily obstructed by foreign matter such as blood, mucus, and dental fragments.

☐ **How is coma defined in children?**

Glasgow Coma or modified Pediatric Coma Score of 8 or less.

☐ **What are the five indications for endotracheal intubation of the child trauma victim?**

1. Respiratory failure/arrest.
2. Airway protection.
3. Airway obstruction.
4. Coma.
5. Need for prolonged ventilatory support or neurological resuscitation.

☐ **Why is the nasotracheal intubation route contraindicated in children less than 8 years of age?**

Because it is extremely difficult to direct the tube anteriorly through the vocal cords in young children without direct visualization. In addition, adenoid tissue is much more prominent and susceptible to injury during emergency intubation.

☐ **What differences make the pediatric airway more difficult than that of an adult?**

Large occiput affects positioning. Larger tongue in small mouth can obstruct view. Larynx is higher (C3-4 vs C6), funnel shaped, narrowest at cricoid ring and more anterior. The epiglottis is relatively large and floppy.

☐ **What are the steps of Rapid Sequence Intubation (RSI)?**

Patient assessment, preparation, resuscitation » Premedication » Preoxygenation » Sedation Agent » Paralyzing agent » Cricoid pressure » Intubation » Confirmation

☐ **What three specific clinical conditions must be recognized prior to selecting sedating agent, and what are the preferred agents for those conditions?**

1. Asthma - Ketamine
2. Increased intracranial pressure - Etomidate
3. Hypotension - Etomidate

☐ **What premedication is given during RSI if you suspect increased ICP?**

Lidocaine, it is optional otherwise to suppress cough reflex.

☐ **Why is atropine given to pediatric patients as a premedication?**

Atropine potentially reduces oral secretions and the risk of laryngoscopy-induced bradycardia. It is frequently used in infants <1yr, children in shock, and if patient is also receiving succinylcholine and/or ketamine.

☐ **T/F: Cricothyrotomy is commonly used to control the pediatric airway.**

False. It is rarely necessary. It should not be performed in children <10yr, needle cricothyrotomy is preferred.

☐ **When may cricothyrotomy be required?**

In the presence of orofacial trauma.

☐ **How should an injured child with adequate ventilation receive supplemental oxygen?**

In the highest available concentration through a nonrebreathing mask.

☐ **What should you do if respiratory effort is ineffective?**

Assist ventilations with a bag-valve-mask device and reservoir delivering 100% oxygen.

☐ **What is associated with respiratory acidosis secondary to injury?**

Alveolar hypoventilation.

☐ **What type of acidosis is caused by hypovolemia and shock?**

Metabolic.

☐ **Why should extreme hyperventilation be avoided?**

It may reduce cerebral blood flow to levels associated with ischemia.

☐ **What factors can compromise ventilation of the injured child?**

Gastric distention and leak around an uncuffed endotracheal tube.

☐ **What is thought to be the leading cause of preventable death in children with multiple injuries?**

Failure to recognize and control internal bleeding.

☐ **Why is blood transfusion of paramount importance in the initial stabilization of the pediatric trauma patient who has sustained significant blood loss?**

To restore oxygen delivery to tissues as well as intravascular volume.

☐ **Bicycle helmets appear to reduce the risk of bicycle-related head injury by about how much?**

80%.

☐ **How do you accomplish the immediate control of external hemorrhage?**

Direct pressure with a gloved hand over the wound using sterile gauze dressings.

☐ **When should tourniquets be used?**

In cases of traumatic amputation or uncontrolled bleeding that cannot be controlled with direct pressure. Damage to skin and ischemia can occur if tourniquet not removed within 2 hours.

☐ **Above what percentage of blood volume loss will signs of shock be observed?**

15%.

☐ **After what percentage of blood loss will hypotension be present?**

After 30% or more of the child's blood volume has been lost acutely.

☐ **What class of hemorrhage is present in compensated shock?**

Class I-II, mild to moderate hypovolemia.

☐ **How should compensated shock be treated?**

With rapid volume replacement with a bolus of 20 ml/kg of isotonic crystalloid solution.

☐ **What two isotonic crystalloids are used in resuscitation?**

Normal saline and lactated ringers.

☐ **What class of hemorrhage is uncompensated shock?**

Class III-IV.

☐ **What is the immediate treatment for uncompensated shock?**

Immediate volume replacement and blood transfusion.

☐ **How should these fluids be administered?**

Using a pressure infusion system or "wide open" intravenous system may be necessary.

☐ **When should urgent transfusion and possibly surgery be considered?**

If the child fails to respond to the administration of two or three boluses of crystalloid solution (approximately 40-60 ml/kg).

☐ **What type of blood transfusion should be given in shock?**

Packed red blood cells mixed with normal saline warmed to body temperature.

☐ **How many times should you repeat these doses?**

Until systemic perfusion is adequate.

☐ **T/F: Administered blood should always be type specific and crossmatched.**

False. Transfusions must not be delayed to await compatibility studies if shock continues despite crystalloid therapy. Instead, O-negative blood should be administered immediately.

☐ **What is indicated if shock persists despite control of external hemorrhage and volume resuscitation?**

Internal bleeding.

TRAUMA

☐ **T/F: Volume resuscitation should be limited in a child with head injury.**

False. Volume resuscitation and blood transfusion should continue as long as signs of shock are present in a child with head injury.

☐ **Won't volume resuscitation increase the likelihood of cerebral injury in head trauma?**

No, ischemia may complicate traumatic brain injury unless intravascular volume is effectively restored.

☐ **T/F: Isolated head injury can cause sufficient blood loss to produce shock in a child.**

True, isolated head injury rarely causes shock but may if bleeding scalp lacerations are not appropriately managed.

☐ **What nasogastric findings support the diagnosis of organ rupture?**

Blood from the NG tube.

☐ **T/F: Serious chest injuries are common in pediatric trauma.**

False, they are uncommon.

☐ **What types of intrathoracic injuries constitute an immediate threat to life.**

Tension pneumothorax.
Open pneumothorax.
Massive hemothorax.
Cardiac tamponade.
Flail chest.

☐ **Why is it more likely to find severe intrathoracic injury without chest wall or rib injury in children than in adults?**

Because the pediatric chest wall is extremely compliant, kinetic energy is transmitted to internal organs with minimal dissipation of the energy occurring.

☐ **What is indicated by the presence of rib fractures?**

That severe chest trauma has occurred, and injury to underlying organs, such as the liver, spleen, and lungs is likely to be present.

☐ **Which two injuries are most likely to impede initial stabilization of the pediatric trauma victim?**

Tension pneumothorax and open pneumothorax.

☐ **How are flail chest and massive hemothorax best managed?**

Initially by aggressive treatment of the respiratory failure and shock they produce.

PEDIATRIC EMERGENCY MEDICINE BOARD REVIEW

☐ **How is flail chest best managed?**

Positive-pressure ventilation.

☐ **How is a hemothorax managed?**

Chest tube.

☐ **T/F: Cardiac tamponade is common in childhood blunt trauma.**

False.

☐ **What causes a tension pneumothorax?**

The trapping of air behind a one-way "flap-valve" or defect in the lung that results from penetrating chest trauma or acute barotrauma sustained at the moment of blunt injury.

☐ **What are the signs of a child with tension pneumothorax?**

Severe respiratory distress.
Distended neck veins.
Contralateral tracheal deviation.
Hyperresonance, decreased chest expansion and diminished breath sounds on the side of injury.
All of these may be difficult to assess in children.

☐ **As tension pneumothorax progresses, what are the effects on systemic circulation?**

Systemic perfusion will be severely compromised as the mediastinum shifts to the contralateral side, twisting the superior and inferior vena cavae and obstructing venous return.

☐ **What is the treatment for tension pneumothorax?**

Needle decompression followed by placement of a chest tube.

☐ **T/F: Tension pneumothorax should be confirmed by chest x-ray prior to decompression.**

False. Decompression must precede confirmatory chest x-ray if signs of respiratory distress or shock are present.

☐ **What type of needle is used for chest decompression?**

14 or 16 gauge angiocatheter.

☐ **Where is the needle inserted?**

Second intercostal space, midclavicular line, just above the third rib.

TRAUMA

☐ **What is another name for open pneumothorax?**

Sucking chest wound.

☐ **What causes open pneumothorax?**

A penetrating chest wound that allows free, bi-directional flow of air between the affected hemithorax and the surrounding atmosphere.

☐ **How does an open pneumothorax prevent effective ventilation?**

By causing equilibration of intrathoracic and extrathoracic pressure. This results in a paradoxical shifting of the mediastinum to the contralateral side with each spontaneous breath. Also, if the defect is 2/3 the diameter of the trachea, air will go through the defect instead of down the trachea.

☐ **Why are sucking chest wounds more lethal in children than adults?**

Because the mediastinum is particularly mobile during childhood.

☐ **What is the primary treatment for respiratory decompensation associated with an open pneumothorax?**

Positive pressure ventilation.

☐ **What else should be done?**

The wound should be covered using an occlusive dressing such as Vaseline® gauze. This dressing should be taped on three sides to allow egress of entrapped air during exhalation. Dressing application should be followed by insertion of a chest tube, unless the defect is so large that it requires immediate surgical repair.

☐ **What is the number one cause of death for African-American males between the ages of 10 and 24?**

Firearm injury.

☐ **Do intentional or unintentional causes account for more firearm-related deaths?**

Intentional, which are thought to account for 94% whereas unintentional firearm injuries account for about 5%.

☐ **What is the order of recruitment of accessory muscles of respiration in distress?**

Subcostal < intercostals < substernal < supraclavicular < nasal flaring < head bobbing.

☐ **What injuries are associated with lap belt or bicycle handlebar injury?**

Perforations of the duodenum and proximal jejunum. Chance fractures are associated with lap belts. Bicycle handlebar injury is frequently the cause of pancreatitis.

PEDIATRIC EMERGENCY MEDICINE BOARD REVIEW

☐ **Facial trauma in children is associated with?**

55 percent of children with facial trauma have associated intracranial injury.

☐ **What bone is most commonly associated with posttraumatic developmental deformities?**

Mandible.

☐ **What is the most common cause of death during the first hour after a burn injury?**

Inhalation injury produces upper airway edema leading to respiratory impairment.

☐ **What is the leading cause of trauma death in children?**

Head injury. Children < 4 years of age have the poorest long term outcomes from severe head injury. This may be related to the higher incident of nonaccidental trauma as a cause for severe injury in this age group.

☐ **What is the most common cause of of head injuries in children?**

Falls.

☐ **What is the primary goal in the management of head injuries in children?**

Prevent secondary insult by preventing hypoxia, ischemia, and increased intracranial pressure.

☐ **Which is worse, decerebrate or decorticate posturing?**

Decerebrate is considered worse as it indicates damage to the midbraine where decorticate posturing signifies damage to the cerebral cortex.

☐ **What is the PECARN rule?**

Pediatric Emergency Care Applied Research Network low-risk clinical decision rule used to determine risk of significant head injury, often used to determine whether or not imaging is necessary.

AGE (YEARS)	CLINICAL CRITERIA
<2	NORMAL MENTAL STATUS NORMAL BEHAVIOR PER CAREGIVER NO LOC (EXCLUDES BRIEF) NO SEVERE MECHANISM NO NON-FRONTAL SCALP HEMATOMA NO EVIDENCE OF SKULL FRACTURE
>2-18	NORMAL MENTAL STATUS NO LOC NO SEVERE MECHANISM NO VOMITING NO SEVERE HEADACHE NO SIGNS OF BASILAR SKULL FRACTURE

TRAUMA

☐ **What is considered a severe mechanism?**

Fall >3ft (<2yr) or >5ft (>2yr); head strike by high impact object; MVC with patient ejection, death to another passenger or rollover; hit by vehicle as pedestrian or bicyclist without helmet

☐ **What are signs of a basilar skull fracture?**

Hemotympanum, CSF rhinorrhea or otorrhea, raccoon eyes, Battle sign

ORTHOPEDIC EMERGENCIES

☐ **A child has posttraumatic tenderness at the end of a long bone. A joint effusion is the only radiographic finding. What is the appropriate management?**

Immobilization and subsequent orthopedic evaluation for possible separation of the epiphysis from the metaphysis (i.e., Salter-Harris type I fracture).

☐ **Describe a common patient with a slipped capital femoral epiphysis.**

Obese boy, 10–16 years old. Groin or knee discomfort increases with activity; they may have a limp. Often bilateral. The slip is best seen on a lateral view.

☐ **What is the most common ankle injury?**

75% of ankle injuries are sprains with 90% of these involving the lateral complex. 90% of lateral ligament injuries are to the anterior talofibular.

☐ **What is the most helpful physical exam test for anterior talofibular ligament injury?**

Anterior drawer test. > 3 mm of excursion might be significant (compare sides); > 1 cm is always significant.

☐ **How are sprains classified?**

1°—stretching of ligament, normal x-ray.
2°—severe stretching with partial tear, marked tenderness, swelling, pain, normal x-ray (now stressed).
3°—complete ligament rupture, marked tenderness, swollen and/or obviously deformed joint. x-ray may show an abnormal joint.

☐ **Which bone is most commonly fractured at birth?**

The clavicle.

☐ **A patient cannot actively abduct her shoulder. What injury does this suggest?**

Rotator cuff tear.

☐ **Why is a displaced supracondylar fracture (of distal humerus) in a child considered a true emergency?**

The injury often results in injury to brachial artery or median nerve and possible compartment syndrome.

☐ **What is the significance of the fat pad sign with an elbow injury?**

Fat pad sign is a radiolucency just anterior to the distal humerus and is indicative of effusion or hemarthrosis of the elbow joint; this suggests an occult fracture of the radial head.

☐ **What type of Salter-Harris fracture has the worst prognosis?**

Type V—compression injury of the epiphyseal plate.

☐ **Which is the more sensitive test used to determine an anterior cruciate ligament tear in the knee: the anterior drawer test or the Lachman test?**

Lachman test. While the knee is held at 20° flexion and the distal femur is stabilized, the lower leg is pulled forward. Greater than 5 mm anterior laxity compared to the other knee is evidence of an anterior cruciate ligament tear.

☐ **What is a stress fracture?**

A fracture is caused by small, repetitive forces usually involving the metatarsal shafts, the distal tibia, and the femoral neck.

☐ **What is "nursemaid's elbow"?**

Subluxation of the radial head. During forceful retraction, some fibers of the annular ligament that encircle the radial neck become trapped between the radial head and the capitellum. On presentation, the arm is held in slight flexion and pronation.

☐ **A patient presents with pain along the radial aspect of the wrist extending into the forearm. What is the diagnostic test of choice?**

Finkelstein's test confirms the diagnosis of deQuervain's tenosynovitis, an overuse inflammation of the extensor pollicis brevis and the abductor pollicis where they pass along the groove of the radial styloid. The patient makes a fist with the thumb tucked inside the other fingers. The test is positive if pain is reproduced when the examiner gently deviates the fist in the ulnar direction.

☐ **In carpal tunnel syndrome, Tinel's sign is produced by tapping the volar wrist over the median nerve. If the test is positive, what does the patient experience?**

Paresthesias extending into the index and long fingers.

☐ **Where is the most common site of compartment syndrome?**

Anterior compartment of the leg.

☐ **What are the most common lower extremity injuries to bone in children?**

Tibial and fibular shaft fractures, usually secondary to twist forces.

ORTHOPEDIC EMERGENCIES

☐ **What radiograph would one order with suspected patellar fracture in a child?**

Standard radiographs including patellar or "sunrise" views, plus comparison radiographs of uninvolved knee.

☐ **What are the differences between avulsion of tibial tubercle and Osgood-Schlatter disease?**

Both occur at tibial tubercle. Avulsion presents with acute inability to walk, lateral view of the knee is most diagnostic; treatment is surgical. Osgood-Schlatter's has vague history of intermittent pain, is bilateral in 25% of cases, has pain with range of motion but not with rest; treatment is symptomatic and not surgical.

☐ **What is a toddler fracture?**

A spiral fracture of the distal 1/3 of the tibia, without fibular involvement.

☐ **What is the most common Salter-Harris class fracture?**

Type II. A triangular fracture involving the metaphysis (Thurston-Holland fragment) and an epiphyseal separation.

☐ **What is the basic disorder contributing to the pathophysiology of compartment syndrome?**

Increased pressure within closed tissue spaces compromising blood flow to muscle and nerve tissue. There are three prerequisites to the development of compartment syndrome:
1. Limiting space.
2. Increased tissue pressure.
3. Decreased tissue perfusion.

☐ **What are the two basic mechanisms for elevated compartment pressure?**

1. External compression—by burn eschar, circumferential casts, dressings, or pneumatic pressure garments.
2. Volume increase within the compartment—hemorrhage into the compartment, IV infiltration, or edema due to post-ischemic (postischial) swelling or secondary to injury.

☐ **What is the most common ligamentous injury to the knee?**

Anterior cruciate ligament.

☐ **What is the weakest portion of a child's bone?**

The physis (epiphyseal plate).

☐ **What is the major concern with injury to the physis?**

The potential for subsequent growth arrest.

PEDIATRIC EMERGENCY MEDICINE BOARD REVIEW

☐ **What is a Type I Salter-Harris fracture?**

A separation of the epiphysis from the metaphysis. Usually caused by avulsion or shearing forces. Prognosis is good.

☐ **What is a Type II Salter-Harris fracture?**

A fracture through the physis which exits the metaphysis. Prognosis is good.

☐ **What is a Type III Salter-Harris fracture?**

A fracture through the physis which exits the epiphysis into the articular surface. Much worse prognosis than Types I & II. ORIF is often required.

☐ **What is a Type IV Salter-Harris fracture?**

A fracture extending through the metaphysis, physis and epiphysis. Carries a high rate of premature physeal arrest.

☐ **What is the most common cause of painful hip in older children?**

Transient synovitis. Also called toxic synovitis, irritable hip, and coxalgia fugax.

☐ **What is a Type V Salter-Harris fracture?**

A crushing injury of the physis. Often difficult to pick up radiographically. Usually diagnosed retrospectively after growth arrest occurs.

☐ **What is a "green stick" fracture?**

A cortical disruption on one side of the bone without a disruption on the other side.

☐ **What are some of the proposed etiologies of acute osteomyelitis in children?**

Bacteremia, trauma, malnutrition, immunocomprised, and recent illness.

☐ **What are some possible findings in a child with a septic hip joint?**

The hip may be held in external rotation and flexed. Any motion of the joint will be very painful.

☐ **What bacterial pathogens are associated with osteomyelitis in children?**

Staph aureus is the most common (40-80% of cases).

☐ **Are x-rays helpful in diagnosing osteomyelitis?**

Periosteal reaction and bony destruction take about 10-14 days to become radiographically evident.

ORTHOPEDIC EMERGENCIES

☐ **What is the treatment of osteomyelitis in children? Of a septic joint?**

Treat with broad-spectrum antibiotics including coverage for S. aureus; neonates should also be covered for group B strep, and gram negative coliforms; treatment for osteomyelitis associated with puncture wounds should provide coverage for Pseudomonas.

☐ **What is the treatment of a septic joint?**

Treatment should provide coverage for S. aureus. Primary treatment for a septic joint is irrigation and drainage of the joint in the operating room, obtaining cultures prior to starting antibiotics. Sexually active patients should be covered for N. gonorrhoeae.

☐ **What is Köhler disease?**

It is an idiopathic avascular necrosis (osteochondrosis) of the tarsal navicular bone. It is unique among the osteochondroses in that it occurs more commonly in school-aged rather than adolescent children.

☐ **Are dislocations and sprains more common in children or adults?**

Dislocations and ligamentous injuries are uncommon in prepubertal children as the ligaments and joints are quite strong as compared to the adjoining growth plates. Excessive force applied to a child's joint is more likely to cause a fracture through the growth plate than a dislocation or sprain.

☐ **Which type of fracture deformity can be particularly difficult to correct?**

Rotational deformities do not correct spontaneously and require meticulous realignment.

☐ **What is a torus fracture?**

A torus, or buckle, fracture is unique to children secondary to compressive forces causing buckling of the bone, leaving the cortices intact.

☐ **What is Volkmann's ischemia?**

It is pain worsened by passive motion of a muscle and associated neurologic deficits secondary to ischemia of the muscles and nerves; it is usually associated with a compartment syndrome.

☐ **What is Volkmann's contracture?**

It is the result of untreated Volkmann's ischemia, a contracture deformity secondary to muscle necrosis.

☐ **What is an antalgic limp?**

It is a painful limp characterized by a shortened stance phase to decrease time on the affected extremity.

☐ **What is a Trendelenburg gait?**

It is a non-painful limp, indicative of underling hip instability or muscle weakness, characterized by an equal stance phase between the involved and uninvolved side. The child will attempt to shift the center of gravity over the involved side for improved balance.

☐ **What are some causes (according to age) of painful limping?**

Antalgic gait may be caused by: (1-3-year-old): infection, occult trauma, or neoplasm; (4-10-year-old): infection, transient synovitis of the hip, Legg-Calve-Perthes disease, rheumatologic disorder, trauma, or neoplasm; (11+-year-old): slipped capital femoral epiphysis, rheumatologic disorder, trauma. It may also be seen in sickle cell disease.

☐ **What are some causes of Trendelenburg gait?**

1-3-year-old: hip dislocation, neuromuscular disease.
4-10-year-old: hip dislocation, neuromuscular disorder.

☐ **What is the Ortolani maneuver?**

Gentle abduction with the hip in flexion while lifting up on the greater trochanter which results in reduction of a dislocated hip with an palpable "clunk" as the femoral head slips over the posterior rim of the acetabulum.

☐ **What is the Barlow maneuver?**

Adduction of the hip in flexion while pushing down gently on the on the femoral head, in a dislocatable hip, will cause dislocation as the femoral head slips out of the acetabulum.

☐ **What is the most sensitive indicator of hip disease?**

Decreased internal rotation of the hip.

☐ **What makes the blood supply to the capital femoral epiphysis vulnerable?**

The retinacular vessels lie on the surface of the femoral neck and are intracapsular; they enter the epiphysis from the periphery and are subject to damage from trauma, joint infection, and other vascular insults.

☐ **What is Legg-Calve-Perthes disease (LCPD)?**

It is an idiopathic avascular necrosis of the capital femoral epiphysis in a child (usually male) aged 2-12 years (mean: 7 years).

☐ **What are some clinical manifestations of LCPD?**

Antalgic gait, muscle spasm (with restricted abduction and internal rotation), proximal thigh atrophy, thigh pain, and mild shortness of stature.

ORTHOPEDIC EMERGENCIES

☐ **What is the most common adolescent hip disorder?**

Slipped capital femoral epiphysis (SCFE).

☐ **What is SCFE?**

It is an inferior and posterior slippage of the head of the femur off the femoral neck. Slips can be either chronic or acute.

☐ **SCFE typically occurs in which individuals?**

Obese adolescents with delayed skeletal maturation or in tall, thin children after a recent growth spurt.

☐ **Is SCFE bilateral or unilateral?**

The disease usually presents unilaterally with thigh, knee, or groin pain and a limp. Up to 30-40% of children with one slip will develop bilateral disease eventually and must be followed for this possibility.

☐ **What radiographic views may be helpful in diagnosing SCFE or LCPD?**

Anteroposterior and Lauenstein (frog-leg) radiographs.

☐ **What is the treatment of SCFE?**

SCFE is one of the few pediatric orthopedic emergencies. Weight bearing must cease to prevent a partial slip from becoming complete (complete slips carry an increased risk of vascular compromise and aseptic necrosis). The treatment is surgical.

☐ **What are the two major complications of SCFE?**

Osteonecrosis (secondary to damage to the retinacular vessels) and chondrolysis (degeneration of the articular cartilage) of the hip.

☐ **What are growing pains?**

They are nighttime thigh and calf pains most commonly seen in 4-10 year olds with symmetric limb pains unaccompanied by systemic symptoms such as fever or weight loss. Although termed growing pains, they are usually not seen in adolescents who undergo a faster rate of growth than younger children.

☐ **Is back pain in children common?**

No, it is unusual and frequently due to organic causes. Etiologies include inflammatory diseases, rheumatologic diseases, developmental diseases, mechanical trauma, and neoplastic diseases.

☐ **What is torticollis?**

Twisted or wry, neck, where the head is tipped to one side, with the occiput rotated toward the shoulder. Etiologies include soft tissue trauma, rotary subluxation of the cervical spine, dystonic reactions, and muscular causes (e.g. viral myositis).

☐ **What is "Little League elbow"?**

Medial epicondylitis or osteochondritis desiccans of the capitellum and/or radial head laterally. Children usually present with pain after throwing or after gymnastics.

☐ **What is osteochondritis desiccans?**

Degenerative process localized to an area of bone and its articular surface that may be due to ischemia or repeated trauma; the exact cause is unknown. The bone becomes avascular and ultimately separates from the underlying proximal bone.

☐ **What is the most common site of osteochondritis desiccans?**

The lateral aspect of the medial femoral condyle; it may also be seen in the ankle and the elbow.

☐ **What is a ganglion?**

It is a fluid-filled cyst at the wrist where a defect in the joint capsule allows herniation of the synovium.

☐ **An adolescent boy presents with pain over the tibial tubercle after athletic activity and presents with a tender prominent tibial tubercle. What is the diagnosis?**

Osgood-Schlatter disease; complete resolution of symptoms over 1-2 years is expected as physeal closure of the tubercle occurs. Symptomatic care with rest, anti-inflammatories, and non-steroidal medication is the usual treatment.

☐ **What is the apprehension test?**

Attempts to displace the patella laterally gives a subjective feeling of imminent patellar subluxation in patients with maltracking patellas who are predisposed to dislocation. The apprehension test refers to the patient's anxiety over this imminent subluxation during the above described maneuver.

☐ **What is the most common carpal bone fractured?**

Scaphoid (navicular).

☐ **What is osteogenesis imperfecta?**

It is the most prevalent of the osteoporotic syndromes in children. It is the result of a variety of defects in the production of type I collagen, with mutations in one of the two genes for type I procollagen. It is divided into four types. Type II is usually lethal in the perinatal period and characterized by "crumpled bone" syndrome and beaded ribs. Approximately 50% are stillborn and most die of respiratory insufficiency. Type III and IV are rare.

ORTHOPEDIC EMERGENCIES

☐ **What are some of the characteristics of Osteogenesis Imperfecta Type I?**

Thin bone cortices, osteoporosis, multiple bone fractures usually occurring between ages 2 and 3 years and continuing through early adolescence. Patients may present with hypermobile joints, kyphoscoliosis, joint dislocations, bowed limbs, increased capillary fragility, thin skin, and blue scleras.

☐ **What is spondylolysis?**

It refers to a defect, lysis, or stress fracture of the pars interarticularis; the pars interarticularis is the bony connection between the superior facet of one vertebra and the posterior elements of the vertebra above it. Symptomatic children may experience low back pain worsened with activity and hamstring tightness.

☐ **What is spondylolisthesis?**

Slippage of the superior vertebra on the lower one. Symptoms of nerve root compression (usually L-5) such as muscle weakness, sphincter problems, and sensory loss may be present.

☐ **What is the order of ossification centers in the elbow?**

Remember the acronym CRITOE. Capitellum, Radial Head, Internal (medial) epicondyle, Trochlea, Olecranon, External (lateral) epicondyle. These ossify ad 3-5-7-9-11-13 years, in that order. Considerable variation to these ages range, but the order of ossification is usually consistent.

☐ **X-ray of the right femur shows a large, dense diaphyseal/metaphyseal lesion with a sun-burst periosteal reaction. The patient's mother reports painful swelling for 3 months with frequent fevers. Probable diagnosis?**

Central osteosarcoma, also consider osteoid osteoma, and sclerosing osteomyelitis.

☐ **Which paraneoplastic syndrome is present in 25% of patients with central osteosarcoma?**

Diabetes mellitus.

☐ **What is the most common malignant bone tumor in children?**

Osteosarcoma is the most common and Ewing's sarcoma is second.

☐ **A 10-year-old boy presents with left hip pain. X-ray shows widening of the epiphyseal plate.**

Slipped capital femoral epiphysis.

☐ **Describe the nature of the injury in slipped capital femoral epiphysis.**

Salter I fracture of the proximal femoral growth plate.

☐ **Who are more likely to be affected?**

Overweight adolescent males. It associated with hypothyroidism, renal osteodystrophy, trauma and growth hormone therapy.

☐ **What findings can be seen on X-rays?**

Blurring of the junction of the metaphysis and epiphysis and widening of the growth plate.

☐ **A 9-year-old boy presents with fever, leukocytosis and right ankle pain. X-rays reveal a central area of lucency surrounded by a rim of reactive sclerosis in the distal tibia and a lucent channel-like configuration extending toward the growth plate. Most common diagnosis?**

Brodie's abscess caused by S. Aureus.

☐ **What are the most common causes of osteomyelitis?**

Skin wounds (S. Aureus) is #1. Genitourinary tract infections, lung infections, and dermal infections are also potential causes.

☐ **What is the most common location of osteomyelitis in children?**

The metaphyses of long bones due to the blood supply.

☐ **X-rays of a 2-year-old demonstrate poorly mineralized epiphyses, widening of the growth plates, fraying and cupping of the metaphyses, and bowing of the long bones. The family are strict vegetarians. Diagnosis?**

Rickets.

EYES, EARS, NOSE, THROAT & NECK

☐ **What is the range of normal intraocular pressure?**

10–20 mmHg. Patients with acute angle closure glaucoma generally have pressures elevated to 40–80 mmHg.

☐ **Topical steroids for the eyes are absolutely contraindicated in what cases?**

When the patient has a herpetic lesion in the eye (often seen as dendritic lesions on slit lamp) or a herpetic infection elsewhere.

☐ **What is the most common finding on funduscopic examination of the AIDS patient?**

Cotton wool spots due to disease of the microvasculature. Other findings are hemorrhage, exudate, or retinal necrosis.

☐ **If you suspect rupture of the round or oval window in a patient that comes to you for acute hearing loss what test might you perform?**

Applying positive pressure to the tympanic membrane will cause the patient to have ipsilateral nystagmus if there is a rupture.

☐ **Acute tinnitus is associated with toxicity of what medication?**

Salicylates.

☐ **Hairy leukoplakia is characteristic of what two viruses?**

HIV and Epstein-Barr virus. Hairy leukoplakia is usually found on the lateral aspect of the tongue. Oral thrush may also be associated with patients with HIV.

☐ **What is the most common cause of odontogenic pain?**

Carious tooth.

☐ **A child falls and knocks out his front tooth. How would treatment differ if the child was age 3 versus age 13?**

With primary teeth, no reimplantation should be attempted because of the risk of ankylosis or fusion to the bone. However, with permanent teeth reimplantation should occur as soon as possible. Remaining periodontal fibers are a key to success. Thus, the tooth should not be wiped dry as this may disrupt the periodontal ligament fibers still attached.

PEDIATRIC EMERGENCY MEDICINE BOARD REVIEW

☐ **What is the best transport medium for an avulsed tooth?**

Hank transports his avulsed permanent tooth in Hank's solution because it has a balanced pH. Hanks > saliva/under tongue > milk > saline > water. Re-implant within 30 minutes.

☐ **What is the most common oral manifestation of AIDS?**

Oropharyngeal thrush.

☐ **A patient complains of a tongue irritation from a slightly chipped tooth after a fall. No dentin is exposed. What treatment can be offered to this patient?**

Tooth fractures only involving the enamel are called Ellis Class I fractures. The sharp edges can be filed with an emery board for immediate relief and the patient can be referred to a dentist for cosmetic repair.

☐ **How can one differentiate Ellis class II and III fractures?**

Class II: Fractures involving the dentin and enamel. The exposed dentin will be pinkish.
Class III: Fractures involving the enamel, dentin, and pulp. A drop of blood is frequently noted in the center of the pink dentin.

☐ **Why should topical analgesics not be used in Ellis class III tooth fractures?**

Severe tissue irritation or sterile abscesses may occur with their use. Treatment includes application of tinfoil, analgesics, and immediate dental referral.

☐ **A patient presents with gingival pain and a foul mouth odor and taste. On exam, fever and lymphadenopathy are present. The gingiva is bright red and the papillae are ulcerated and covered with a gray membrane. What is the diagnosis and treatment?**

Acute necrotizing ulcerative gingivitis (ANUG). Can result in soft tissue and alveolar bone necrosis. Treat with tetracycline/doxycycline or penicillin, warm saline rinses, dilute H_2O_2, and viscus lidocaine.

☐ **A 17-year-old female presents complaining of excruciating waxing and waning "electric shock type" pain in the right cheek. What is the diagnosis and treatment?**

Tic douloureux. The most significant finding is that the pain follows the distribution of the trigeminal nerve. Carbamazepine (100 mg bid starting dose and increasing to 1200 mg daily if needed).

☐ **A three-year-old child presents with a unilateral purulent rhinorrhea. What is the most likely diagnosis?**

Nasal foreign body.

☐ **What potential complications of nasal fractures should always be considered on physical examination?**

Septal hematoma and cribriform plate fractures. A septal hematoma appears as a bluish mass on the nasal septum and if not drained aseptic necrosis of the septal cartilage and septal abnormalities may occur. A cribriform plate fracture should be considered in a patient who has a clear rhinorrhea after trauma.

EYES, EARS, NOSE, THROAT AND NECK

☐ **What physical exam findings would make posterior epistaxis more likely than anterior epistaxis?**

1. Inability to see the site of bleeding. Anterior nosebleeds usually originate at Kiesselbach's plexus, an area easily visualized on the nasal septum.
2. Blood from both sides of the nose. In a posterior nosebleed the blood can more easily pass to the other side because of the proximity of the choanae.
3. Blood trickling down the oropharynx.
4. Inability to control bleeding by direct pressure.
5. Posterior bleed is rare in children.

☐ **What is the most common site of bleeding in posterior nosebleeds?**

The sphenopalatine artery's lateral nasal branch.

☐ **A patient returns to the emergency department with fever, nausea, vomiting and hypotension, two days after having nasal packing placed for an anterior nosebleed. What potential complication of nasal packing should be considered?**

Toxic shock syndrome.

☐ **A child with a sinus infection presents with proptosis, a red swollen eyelid and a inferolaterally displaced globe. What is the diagnosis?**

Orbital cellulitis and abscess associated with ethmoid sinusitis.

☐ **A patient with a frontal sinusitis presents with a large forehead abscess. What is the diagnosis?**

Pott's puffy tumor. This is a complication of frontal sinusitis where the anterior table of the skull is destroyed allowing the formation of the abscess.

☐ **An ill-appearing patient presents with a fever of 103°F, bilateral chemosis, third nerve palsies, and untreated sinusitis. What is the diagnosis?**

Cavernous sinus thrombosis. This life-threatening complication occurs from direct extension through the valveless veins. Complication of sinusitis may be local (osteomyelitis), orbital (cellulitis), or within the central nervous system (meningitis or brain abscess).

☐ **Retropharyngeal abscess is most common in what age group, and why?**

Six months to three years. This is due to size regression of the retropharyngeal lymph nodes after the age of three.

☐ **How do patients with retropharyngeal abscesses appear?**

Febrile, ill-appearing, stridorous, drooling, and in an opisthotonic position. These children may complain of difficulty swallowing, or may refuse to feed.

PEDIATRIC EMERGENCY MEDICINE BOARD REVIEW

☐ **What radiographic sign can be used to make the diagnosis of a retropharyngeal abscess?**

A widening of the retropharyngeal space (normal 3-4 mm or less than half the width of the vertebral bodies). False widening may occur if the x-ray is not done during inspiration with the patient's neck extended (occasionally, an air fluid level may be noted in the retropharyngeal space).

☐ **What is the most common organism to cause a retropharyngeal abscess?**

ß-hemolytic Streptococcus.

☐ **A 15-year-old male presents with a high fever, trismus, dysphagia, and swelling inferior to the mandible in the lateral neck. What is the diagnosis?**

Parapharyngeal abscess.

☐ **What is the most common origin of Ludwig's angina?**

The many S's of Ludwig's angina:
Second and third molar origin.
Swelling of Submandibular, Sublingual, Submaxillary Spaces displacing tongue upward and posteriorly.
Strep and Staph and mixed anaerobes and aerobes.
Support airway.

☐ **A 16-year-old female presents with dull right ear and jaw pain and a burning sensation in the roof of her mouth, which is worse in the evening. She also hears a "popping" sound when she opens and closes her mouth. Exam reveals tenderness of the joint capsule. What is the diagnosis and treatment?**

TMJ syndrome. Treat with physiotherapy, analgesia, soft diet, muscle relaxants and occlusive therapy. Apply warm moist compresses 4–5 times daily for 15 minutes for 7–10 days.

☐ **A 14-year-old diabetic male presents with pain, itching, and discharge from the right ear. The tympanic membrane is intact. What is the diagnosis?**

Otitis externa. Treat by suctioning ear and 1 week of antibiotic steroid otic solution. An ear wick may improve delivery of the antibiotic. Suspect malignant otitis externa in the diabetic patient.

☐ **A patient presents with ear pain and fluid-filled blisters on the tympanic membrane. What is the diagnosis?**

Bullous myringitis, commonly caused by Mycoplasma or viruses. Treat with erythromycin.

☐ **A patient presents with a swollen, tender, red left auricle. What is the diagnosis?**

Perichondritis caused by Pseudomonas.

☐ **What is the most frequent cause of hearing loss?**

Cerumen impaction.

EYES, EARS, NOSE, THROAT AND NECK

☐ **If a patient has bilateral sensory hearing loss, what causes should be suspected?**

Idiopathic > Infection > Otologic disease > Trauma > Vascular/Heme > Neoplastic > Other

☐ **Name some causes of tympanic membrane perforation.**

Blast injuries (water or air), foreign bodies in the ear (particularly Q-tips), lightning strikes, otitis media, associated temporal bone fractures.

☐ **What is the most common organism which causes pediatric acute otitis media?**

Streptococcus pneumoniae, followed by Haemophilus influenzae and Moraxella catarrhalis.

☐ **What is a Bezold abscess?**

A complication of acute mastoiditis where infection spreads to the soft tissues below the ear and sternocleidomastoid muscle.

☐ **What is the most common cause of sialoadenitis?**

Mumps.

☐ **A patient presents with trismus, fever, and an erythematous, tender parotid gland. Pus is expressed from Stensen's duct. What conditions may predispose a person to bacterial parotitis?**

Any situation which decreases salivary flow. Irradiation, phenothiazine, antihistamines, parasympathetic inhibitors, dehydration or debilitation. Up to 30% of cases occur postoperatively.

☐ **In which salivary gland do stones most likely occur?**

80% are submandibular.

☐ **What does grunting versus inspiratory stridor indicate?**

Grunting is specific for lower respiratory tract disease, such as pneumonia, asthma, or bronchiolitis. Stridor localizes partial respiratory obstruction at or above the larynx.

☐ **What is the most common cause of epiglottitis?**

Group A Strep, S. aureus, Strep pneumoniae now more common than Hib.

☐ **A lower airway foreign body is suspected. What will plain films show?**

Unilateral hyperinflation. Mediastinal shift AWAY from affected side.

PEDIATRIC EMERGENCY MEDICINE BOARD REVIEW

☐ **A patient presents with an itching, tearing, right eye. On exam huge cobblestone papillae are found under the upper lid. What is the diagnosis?**

Allergic conjunctivitis.

☐ **A patient is seen with herpetic lesions on the tip of the nose. Why is this a problem?**

The tip of the nose and the cornea are both supplied by the nasociliary nerve. Thus, the cornea may also be involved.

☐ **A patient presents with inflammation of the conjunctiva and lid margins. Slit lamp exam reveals a "greasy" appearance of the lid margins with scaling, especially around the base of the lashes. What is the diagnosis?**

Blepharitis. Often caused by staphylococcal infection of the oil glands and skin next to the lash follicles. Treatment consists of scrubbing with baby shampoo, and, in consultation with an ophthalmologist, sulfacetamide drops and steroids.

☐ **A patient presents with a pustular vesicle at the lid margin. What is the diagnosis and treatment?**

Hordeolum (stye). An acute inflammation of the meibomian gland most commonly of the upper lid. Treat with topical antibiotics and warm compresses. Surgical drainage may be necessary.

☐ **A patient presents with a chronic non-tender, uninflamed nodule of the upper lid. What is the diagnosis?**

Chalazion.

☐ **A patient presents with the sensation of a foreign body in the eye. Slit-lamp reveals a dendritic figure which has a Christmas tree pattern. What is the appropriate treatment?**

Suspect HSV Keratitis. Treat with topical antiviral and cycloplegic.

☐ **What are complications of a hyphema?**

The 4 S's.
1. Secondary rebleeds, which usually occur between the second and fifth days post injury since this is the time of clot retraction. Rebleeds tend to be worse than the initial bleed.
2. Significantly increased intraocular pressure which can lead to acute glaucoma, chronic late glaucoma, and optic atrophy.
3. Staining of the cornea due to hemosiderin deposits.
4. Synechiae which interfere with iris function.

☐ **Why do patients with sickle cell anemia and a hyphema require special consideration when presenting with ophthalmologic concerns?**

Increased intraocular pressure can occur if the cells sickle in the trabecular network, preventing aqueous humor from leaving the anterior chamber. Medication, such as hyperosmotics and Diamox, which increase the likelihood of sickling, must be avoided.

EYES, EARS, NOSE, THROAT AND NECK

☐ **A patient presents with a history of trauma to the orbit with dull ocular pain, decreased visual acuity, and photophobia. Exam reveals a constricted pupil and ciliary flush. What will be found on a slit-lamp exam?**

Flare and cells in the anterior chamber are present with traumatic iritis.

☐ **What are causes of a subluxed or dislocated lens?**

Trauma, Marfan's syndrome, homocystinuria, Weill-Marchesani syndrome.

☐ **Physiologically, what causes flare?**

Flare is caused by inflammatory proteins resulting in the "dust in the movie projector lights" or "fog in the headlights" phenomena in slit-lamp examination.

☐ **Why shouldn't topical ophthalmologic anesthetics be prescribed?**

They inhibit wound healing and decrease patients' sensation, predisposing to injury. The evidence for this answer is very weak.

☐ **What is the most common organism in contact lens associated corneal ulcers?**

Pseudomonas.

☐ **How can Krazy-Glue (cyanoacrylate) be removed if a patient has stuck their eyelids together?**

If eyes sealed in normal position, use gentle traction first otherwise let glue dissolve naturally over time. Consult optho if eyes glued inverted. Avoid use of other substances to dissolve. Treat co-occurring corneal abrasions.

☐ **What test can a physician use to determine if blindness is of a hysterical origin?**

An optokinetic drum can be used. If nystagmus eye movements occur, the patient is seeing the stripes. Prisms may also be used.

☐ **A patient presents with traumatic severe pain behind the left eye, a left pupil afferent defect, central visual loss, and a left swollen disc. What is the diagnosis and potential causes?**

Optic neuritis. This may be idiopathic or may be associated with multiple sclerosis, Lyme disease, neurosyphilis, lupus, sarcoid, alcoholism, toxins, or drug abuse.

☐ **In what age group is retropharyngeal abscess most common?**

50% cases occur in 6-12 months of age. 96% all cases seen in <6 years of age. Look for difficulty breathing, fever, cervical adenopathy, dysphagia, stiff neck.

☐ **In what age group is a peritonsillar abscess most common?**

Adolescents and young adults. Look for ear pain, trismus, drooling, voice change. Treat with antibiotics and I&D or needle aspiration.

PEDIATRIC EMERGENCY MEDICINE BOARD REVIEW

☐ **What is the most common sight-threatening ocular infection in AIDS patients?**

Cytomegalovirus (CMV) retinitis occurs in 15-40% of AIDS patients.

☐ **What are the ocular manifestations of Lyme disease?**

Uveitis, keratitis, and optic neuritis.

☐ **Extended wear contact lenses worn overnight increase the risk of infectious keratitis by how much compared to daily wear?**

Approximately 10-fold.

☐ **What type of conjunctivitis is caused by adenovirus?**

Acute follicular conjunctivitis.

☐ **What is the most common cause of ophthalmia neonatorum in the U.S.?**

Chlamydia trachomatis.

☐ **What diagnosis is most probable in a newborn who develops "hyperacute" conjunctivitis with copious mucopurulent discharge?**

Neisseria gonorrhea.

☐ **A patient with chronic follicular conjunctivitis is found to have several smooth, centrally umbilicated papules on the eyelids. What is the most likely diagnosis?**

Molluscum contagiosum.

☐ **]What two organisms are most likely to cause endogenous bacterial endophthalmitis in a host not otherwise predisposed to infection?**

Neisseria meningitidis and Haemophilus influenzae

☐ **What is the most frequent condition associated with orbital cellulitis?**

Sinusitis.

☐ **How common is strabismus during the first few months of life?**

Occurs in 1/3 of infants < 6 months.

☐ **What is the big deal about having a button battery lodge in your esophagus?**

Pressure necrosis, liquefactive necrosis from alkaline leakage, perforation, tissue electrolysis.

PSYCHOSOCIAL AND CHILD ABUSE

☐ **What is the ratio of attempted suicides to completed suicides?**

10:1.

☐ **Is violence more likely between family members or non-family members?**

Between family members. 20–50% of murders in the US are committed by members of the victims' families. Spouse abuse is as high as 16% in the US.

☐ **In what percentage of cases of child sexual abuse is the abuser known by the child?**

90%.

☐ **In what percentage of cases of child abuse is the mother also abused?**

50%.

☐ **What are the most common ages for child physical abuse?**

Two-thirds of physically abused children are under the age of 3, and one-third are under the age of 6 months.

☐ **In addition to an evaluation for child abuse, what laboratory studies should be done in the child with multiple bruises?**

CBC with differential, platelet count, PT, and UA.

☐ **What is the most common cause of referral to a child psychiatrists?**

Attention Deficit Hyperactivity Disorder.

☐ **What are the side-effects of Ritalin?**

Ritalin (Methylphenidate) is a psychostimulant used to treat ADHD. Side-effects include depression, headache, hypertension, insomnia, and abdominal pain.

☐ **What is the most frequent first episode of bipolar disease, mania or depression?**

Mania. Depression is rarely the first symptom. In fact, only 5–10% of patients that develop depression first go on to have manic episodes. 1/3 of manic patients never have a depressive episode.

PEDIATRIC EMERGENCY MEDICINE BOARD REVIEW

☐ **Other than the classic mania, what can lithium be used to treat?**

Bulimia, anorexia nervosa, alcoholism in patients with mood disorders, leucocytosis in patients on antineoplastic medication, cluster headaches, and migraine headaches.

☐ **Should people who are physically active and taking lithium have their lithium dosage increased or decreased?**

Increased. Lithium, a salt, is excreted more than sodium during sweating.

☐ **T/F: A patient starting lithium will be expected to gain weight.**

True. All psychotropic medications cause weight gain, hence lithium's usefulness in treating anorexia nervosa.

☐ **Lithium toxicity begins at what level?**

14 mg/L. Above this level nausea, diarrhea, vomiting, rigidity, tremor, ataxia, seizures, delirium, coma, and death can all occur.

☐ **Why might a patient taking lithium experience polyuria?**

Long term lithium ingestion can cause nephrogenic diabetes.

☐ **What is the most common clinical symptom of a patient with a borderline personality disorder?**

Chronic boredom. Other symptoms include severe mood swings, volatile relationships, continuous and uncontrollable anger, and impulsiveness.

☐ **A patient presents to your office with parotid gland swelling and erosion of the enamel on her teeth. What findings might you expect to find in this patient?**

The patient described most likely has bulimia. Elevated serum amylase and hypokalemia are associated with bulimia.

☐ **List some common laboratory findings associated with eating disorders?**

Hyponatremia, hypokalemia, hypocalcemia, hypophosphatemia, anemia, hypoglycemia, starvation ketoacidosis, abnormal glucose tolerance, hypothyroidism due to low T3 levels, persistently elevated cortisol due to starvation, low FSH, LH and estrogens, and elevated growth hormone.

☐ **What is the prevalence of conduct disorder?**

10%. It is much more common in boys and is familial.

☐ **Children with conduct disorder are most likely to develop what adult disorder?**

About 40% will have some pathology as adults. The most common disorder is antisocial personality disorder.

PSYCHOSOCIAL AND CHILD ABUSE

☐ **What is conversion disorder?**

Internal psychological conflict that manifests itself as somatic symptoms. Voluntary motor or sensory function is affected. Examples include weakness, imbalance, dysphagia, and changes in vision, hearing, or sensation. These symptoms are not feigned or intentionally produced. They are also not fully explained by medical conditions.

☐ **Delirium is most likely to occur in what kinds of patients?**

Those with multiple medical problems, decreased renal function, high WBC count, anticholinergic, propranolol, scopolamine, or flurazepam drug use.

☐ **Describe delirium.**

"Clouding of consciousness" resulting in disorientation, decreased alertness, and impaired cognitive function. Acute onset, visual hallucinosis, and fluctuating psychomotor activity are all commonly seen. All symptoms are variable and may change over hours.

☐ **Name some over-the-counter and "street" drugs that may produce delirium or acute psychosis.**

Salicylates, antihistamines, anticholinergics, alcohols, phencyclidine, LSD, mescaline, cocaine, and amphetamines.

☐ **Name some symptoms of major depression.**

IN SAD CAGES:
Interest
Sleep
Appetite
Depressed mood
Concentration
Activity
Guilt
Energy
Suicide

☐ **What is dysthymia?**

Dysthymia is a chronic disorder lasting more than 2 years. The severe symptoms of depression, such as delusions and hallucinations are absent. Patients with dysthymia have some good days, they react to their environment, and they have no vegetative signs. 10% of patients with dysthymia develop major depression.

☐ **Wild and abundant dreams may result from withdrawal of what drug?**

Antidepressants. Other side-effects of withdrawal are anxiety, akathisia, bradykinesia, mania, and malaise.

PEDIATRIC EMERGENCY MEDICINE BOARD REVIEW

☐ **What is a dystonic reaction?**

It is a very common side-effect of neuroleptics. Muscle spasm of tongue, face, neck, and back are seen. Severe laryngospasm and extraocular muscle spasm may occur also.

☐ **Define the following dyspraxias: ideomotor, kinesthetic, and constructional.**

Ideomotor: The patient is unable to perform simple motor tasks despite adequate understanding, sensory and motor strength.
Kinesthetic: The patient is unable to position his extremities on command despite adequate understanding, sensory and motor strength.
Constructional: The patient cannot copy simple shapes despite adequate understanding, sensory and motor strength.

☐ **A 16-year-old male presents to the ED complaining of pleuritic pain, palpitations, dyspnea, dizziness and tingling in his arms and legs. What is the diagnosis?**

Hyperventilation syndrome. This is frequently associated with anxiety. The tingling is due to decreased carbonate in the blood.

☐ **In olfactory hallucinations, is the olfactory sensation still intact?**

Yes. Patients remain responsive to normal odors. Hallucinations of foul odors are superfluous to normal smells.

☐ **A 17-year-old male arrives somnolent with vitals of P: 130, R: 26, BP: 170/80, and T: 105°F. You note diffuse muscular rigidity and intermittent focal muscle twitching/jerking lasting 1–2 seconds. As you "work him up," your nurse returns from the waiting area with news from the family that the patient has had a progressive decline of mental status for 2 days after seeing his psychiatrist. The patient has had a history of 'psychosis' for almost one year. What process should be included in your differential diagnosis at this time?**

Neuroleptic malignant syndrome.

☐ **What labs would you expect to be elevated in this patient?**

CPK may elevated and correlates with increasing risk of fatality due to myoglobinuria. Serum alkaline phosphatase, serum aminotransferases and the white cell count are elevated. Hyponatremia, and hypokalemia are also present.

☐ **How is lethal catatonia differentiated from neuroleptic malignant syndrome (NMS)?**

They are differentiated by the timing of the hyperthermia. In lethal catatonia, severe hyperthermia occurs during the excitement phase before catatonic features develop. In neuroleptic malignant syndrome, hyperthermia develops later in the course of the disease with the onset of stupor.

☐ **What is the difference between malingering and factitious disorder?**

The goal of a malingerer is an external incentive such as workman's comp. The goal of a someone with factitious disorder is to enter into the sick role. Both involve voluntary faking of an illness.

PSYCHOSOCIAL AND CHILD ABUSE

☐ **What is an extreme case of factitious disorder?**

Munchausen's syndrome. These patients may actually try to cause harm to themselves (i.e., by injecting feces into their veins) and are very accepting/seeking of invasive procedures. Munchausen by proxy is another example. In this disease the patient seeks medical care for another, usually a child.

☐ **A patient has ingested a phenothiazine and arrives hypotensive. What intervention(s) may be considered?**

IV crystalloid boluses usually suffice. Severe cases are best managed with norepinephrine (Levophed), or metaraminol (Aramine). These pressors stimulate Alpha-adrenergic receptors preferentially. Beta-agonists, such as isoproterenol (Isuprel), are contraindicated due to the risks of Alpha-receptor stimulated vasodilation.

☐ **What are extrapyramidal reactions?**

Dopamine receptor blockade within the nigrostriatal system results in involuntary and spontaneous motor responses including dystonia, akathisia, and Parkinson-like syndrome.

☐ **A 14-year-old female comes to your office complaining of sudden episodes of palpitations, diaphoresis, lightheadedness, a fear of losing control, a sense of being choked, tremors, and paresthesias. What is the diagnosis?**

Panic disorder. Panic disorders need not be linked to any events. Though they are commonly associated with agoraphobia, social phobia, mitral prolapse, and late non-melancholic depression.

☐ **What is the most common specific phobia exhibited during childhood?**

Animal phobia.

☐ **A 16-year-old female was raped 6 months ago. She has now suddenly developed recurrent, distressing flashbacks of the event; she is having nightmares and intense fear, is avoiding all males. She has a diminished memory of the rape, and an exaggerated startle response. Is this woman experiencing PTSD?**

Yes. This is delayed onset PTSD. The onset of symptoms can occur at least 6 months after the provoking event.

☐ **What are the signs and symptoms suggestive of an organic source of psychosis?**

Acute onset, disorientation, visual or tactile hallucinations, age less than 10, and evidence suggesting overdose or acute ingestion, such as abnormal vital signs, pupil size and reactivity, or nystagmus.

☐ **What is the difference between schizophrenia and schizophreniform disorder?**

Schizophreniform disorder implies the same signs and symptoms as schizophrenia, yet these symptoms have been present for less than 6 months. The impaired functioning in schizophreniform disorder is not consistent. Schizophreniform disorder is generally a provisional diagnosis with schizophrenia following.

PEDIATRIC EMERGENCY MEDICINE BOARD REVIEW

☐ **What are the 5 criteria for a diagnosis of Schizophrenia?**

1. Psychosis.
2. Emotional blunting.
3. No affective features or episodes.
4. Clear consciousness.
5. The absence of coarse brain disease, systemic illness, and drug abuse.

☐ **Separation anxiety has an average onset of what age?**

Age 9. These children fear leaving home, sleep, being alone, going to school and losing their parents. 75% develop somatic complaints in order to avoid attending school.

☐ **A 15-year-old female complains of pain in her calf, headache, shooting pain when flexing her right wrist, random epigastric pain, bloating, and irregular menses, all of which cannot be explained following medical workup. What is the diagnosis?**

Somatization Disorder—multiple, unexplained medical symptoms involving multiple systems. In order to diagnose a patient with somatization disorder, one must have 4 or more unexplained pain symptoms. Symptoms generally begin in childhood and are full blown by the age of 30. This is more common in females than males.

☐ **What are some important clinical considerations when handling intoxicated, violent, psychotic, or threatening patients?**

These patients require a careful history and physical with attention to the mental status exam. Look for evidence of trauma, toxic ingestion, or metabolic derangement. Historical sources, such as family, paramedics, mental health workers, police, or medical records may need to be accessed. The patient may need to be physically or chemically restrained to obtain an adequate examination and for the safety of the patient and hospital staff.

☐ **What psychiatric problems are associated with violence?**

Acute schizophrenia, paranoid ideation, catatonic excitation, mania, borderline and antisocial personality disorders, delusional depression, posttraumatic stress disorder, and decompensating obsessive/compulsive disorder.

☐ **What are the prodromes of violent behavior?**

Anxiety, defensiveness, volatility, and physical aggression.

☐ **Define the following:**

Akathisia: Internal restlessness. The patient feels as if he is "jumping out of his skin."
Echolalia: Meaningless automatic repetition of someone else's words. This may occur immediately or even months after hearing the words.
Catalepsy: The patient maintains the same posture over a long period of time.
Waxy flexibility: The patient offers resistance to anyone trying to change his position, then gradually allows himself to be moved to a new posture, much like a clay figure.

PSYCHOSOCIAL AND CHILD ABUSE

☐ **What are the two subclassifications of anorexia nervosa?**

Restricting Type and Binge-Eating/Purging Type

☐ **How do you diagnose bulimia nervosa?**

A history of overeating followed by forged purging is all that is needed. Erosion of the tooth enamel and back of second and third fingers in the dominant hand should lead you to suspect the diagnosis.

☐ **What electrolyte abnormalities are commonly seen in bulimia nervosa?**

Hypokalemia, hypochloremia and metabolic alkalosis.

☐ **What are the main causes of death in patients with anorexia nervosa?**

About 10% of these patients end up dying from complications (i.e. electrolyte imbalances, arrhythmias, or congestive heart failure).

☐ **How long after intake can marijuana still be detected in the urine?**

From 3 days after one episode to 28 days in a frequent user.

☐ **In addition to the history, physical, laboratory tests, and collection of physical evidence, what else needs to be done in a case of sexual abuse of a child?**

File a report with child protective services and law enforcement agencies. Provide emotional support to the child and family. Give a return appointment for follow-up of STD cultures and testing for pregnancy, HIV, or syphilis, as indicated. Assure follow-up for psychological counseling by connecting the child/family to the appropriate services in your area.

☐ **X-ray of a 6-month-old mentally retarded boy shows multiple fractures of varying age and location. Diagnosis?**

Child abuse.

☐ **What long bone fractures may be seen in abused children?**

- Multiple fractures of varying age and location.
- Avulsion fractures of ligamentous insertions.
- Corner fractures.
- Isolated spiral fractures.
- Common locations include multiple ribs, sternum, vertebral compressions, tibia costochondral or costovertebral separation, clavicles, scapula, skull, and metacarpus.

☐ **What is the most common cause of death in abused children?**

Head trauma.

☐ **What are the neuroradiologic findings in child abuse?**

- Subdural hematoma is the most common finding.
- Contusions.
- Cerebral edema.
- Subarachnoid hemorrhage.
- White matter shear injuries at the corticomedullary junction, centrum semiovale, and corpus callosum.
- Epidural hemorrhage.

☐ **What are some typical findings of visceral trauma in child abuse?**

Duodenal hematoma, organ contusion, bowel or gastric rupture, and traumatic pancreatic pseudocysts. Visceral trauma is the second most common cause of death in child abuse.

OBSTETRICS AND GYNECOLOGY

☐ **What is the most common non-gynecologic condition that presents with lower abdominal pain?**

Appendicitis.

☐ **What is secondary amenorrhea?**

No menstruation for 6 months or more in a women who previously had regular menses.

☐ **What is the most common cause of secondary amenorrhea?**

Pregnancy. The second most common cause is hypothalamic hypogonadism which can be due to weight loss, anorexia nervosa, stress, excessive exercise, or hypothalamic disease.

☐ **What is the most common cause of dysfunctional uterine bleeding in a child?**

Infection. For teenagers, it is most commonly due to anovulation.

☐ **Virtually all severe cases of dysfunctional uterine bleeding occur in what age group?**

Adolescent females shortly after the onset of menstruation.

☐ **What is the treatment of choice for a Bartholin gland abscess?**

Marsupialization. This prevents recurrences.

☐ **B-hCG is secreted by what and for what reason?**

B-hCG is secreted by placental trophoblasts; its purpose is to maintain the corpus luteum, which in turn maintains the uterine lining. The corpus luteum is maintained through the sixth to eighth week of pregnancy, by which time the placenta begins to produce its own progesterone for maintenance of the endometrium.

☐ **Failure of the hCG to increase by 66% over 48 hours in a pregnant females is indicative of what?**

An ectopic pregnancy or an intrauterine pregnancy that is about to abort.

☐ **What is the most common site of implantation in an ectopic pregnancy?**

The ampulla of the fallopian tube (95%).

PEDIATRIC EMERGENCY MEDICINE BOARD REVIEW

☐ **What are the risk factors for ectopic pregnancy?**

Prior scarring of the fallopian tubes from infection (i.e., PID or salpingitis), IUDs, previous ectopic pregnancy, tubal ligation, and STDs.

☐ **What percentage of pregnancies are ectopic?**

1.5%. Ectopic pregnancies are the leading cause of death in the first trimester.

☐ **When does an ectopic pregnancy most commonly present?**

6–8 weeks. Patients usually present with amenorrhea and sharp abdominal or pelvic pain. A mass can be felt in 1/2 of the patients. B-hCG confirms pregnancy; ultrasound generally confirms an ectopic pregnancy.

☐ **What is the most common finding on pelvic exam in a patient with an ectopic pregnancy?**

Unilateral adnexal tenderness.

☐ **Who is eligible for methotrexate treatment of an ectopic pregnancy?**

Patients who are hemodynamically stable with unruptured gestations of < 4 cm in diameter by ultrasound.

☐ **Transient pelvic pain frequently occurs 3–7 days after methotrexate therapy for ectopic pregnancy. What is your concern?**

Pain from rupture may be difficult to differentiate from pain presumed to be due to tubal abortion, which typically lasts 4–12 hours.

☐ **For a gestational sac to be visible on ultrasound, what must the B-hCG be?**

At least 6,500 mIU/ml for a transabdominal ultrasound and 2,000 mIU/ml for a transvaginal ultrasound.

☐ **When can an abdominal ultrasound find an intrauterine gestational sac?**

In the fifth week. Fetal pole, sixth week. Embryonic mass with cardiac motion, seventh week.

☐ **Matching—time to keep those "M"s straight:**

1. Menorrhagia. a. Bleeding between menstrual periods.
2. Metrorrhagia (hypermenorrhea). b. Excessive amount of blood or duration.
3. Menometrorrhagia. c. Excessive amount of blood at irregular frequencies.

Answers: (1) b, (2) a, and (3) c.

OBSTETRICS AND GYNECOLOGY

☐ **Matching—more ways to describe menstrual bleeding:**

1. Hypermenorrhea.   a. Menstrual periods > 35 days apart.
2. Oligomenorrhea.   b. Menstrual periods < 21 days apart.
3. Polymenorrhea.   c. Menorrhagia.

Answers: (1) c, (2) a, and (3) b.

☐ **What is the most common complication of ovarian cysts?**

Torsion of the ovary. Torsion is more common with small to medium sized tumors. Emergency surgery is required.

☐ **Most cases of PID are caused by what 2 organisms?**

Neisseria gonorrhoeae and Chlamydia trachomatis.

☐ **What predisposes a female to yeast infections?**

Diabetes, oral contraceptives, and antibiotics.

☐ **Which patients with PID should be admitted?**

Admit patients who are pregnant, have a temperature > 38°C (100.4°F), are nauseous or vomiting (which prohibits po antibiotics), have pyosalpinx or tubo-ovarian abscess peritoneal signs, have IUCD, show no response to oral antibiotics, or for whom diagnosis is uncertain.

☐ **What are the criteria for diagnosis of PID?**

All of the following must be present: (1) adnexal tenderness, (2) cervical and uterine tenderness, and (3) abdominal tenderness. In addition, one of the following must be present: (1) temperature > 38°C, (2) endocervix Gram's stain positive for Gram-negative intracellular diplococci, (3) leukocytosis > 10,000 /mm3, (4) inflammatory mass on US or pelvic, or (5) WBCs and bacteria in the peritoneal fluid.

☐ **Teenage pregnancies are associated with increased risks of what disorders?**

Maternal complications, such as gonorrhea, syphilis, toxemia, anemia, malnutrition, low birth weight babies, and perinatal mortality.

☐ **Why is Rh status important in a pregnant patient?**

If the mother is Rh negative and the fetus is Rh positive, fetal anemia, hydrops, and fetal loss can result. Rh immunoglobulin should be given to all Rh negative patients.

☐ **What organism is the most common cause of vaginitis?**

Candida albicans.

PEDIATRIC EMERGENCY MEDICINE BOARD REVIEW

☐ **A patient presents with a 2 day history of vaginal itching and burning. On exam, you note a thin yellowish–green bubbly discharge and petechiae on the cervix (also known as "strawberry cervix"). What test do you perform, and what results do you expect to find?**

Look at the discharge mixed with saline under a microscope. If you see a mobile and pear-shaped protozoa (Trichomonas vaginalis) with flagella, then the patient should be treated with Metronidazole (Flagyl).

☐ **A 16-year-old sexually active female is complaining of a heavy but thin discharge with an unpleasant odor. When you add 10% KOH to the discharge, the combination emits a fishy odor. What would you expect to see on microscopic examination?**

"Clue cells"—epithelial cells with bacilli attached to their surfaces. This patient has Gardnerella vaginitis. The patient and her partner should both be treated with Metronidazole (Flagyl).

☐ **When should you avoid using Flagyl to treat a female?**

If she is the first trimester of pregnancy. Metronidazole may have teratogenic effects. Clotrimazole (Gyne-Lotrimin) may be used instead. Flagyl has side-effects including nausea, vomiting, and metallic tastes. Most importantly, it acts similarly to disulfiram (Antabuse) and should, therefore, not be taken with alcohol.

☐ **A 15-year-old female presents to the ED complaining of a painful sore on her vulva that looks like a pimple at first. On exam, you find an ulcer with vague borders and a gray base. What is the most likely diagnosis?**

Gram's stain, culture, and biopsy (used in combination because of the high false negative rates) should show that this woman has Haemophilus ducreyi, causing chancroid. Treatment is erythromycin or ceftriaxone.

☐ **Condylomata acuminata frequently occur in combination with what other STD?**

Trichomonas vaginitis.

☐ **What is the most common cause of septic arthritis in young adults?**

Disseminated gonococcal infection.

☐ **What is the treatment for gonorrhea?**

Ceftriaxone and doxycycline. The latter is given because half of the patients infected with gonorrhea are simultaneously infected with chlamydia.

☐ **What is the predominant organism in a healthy female's vaginal discharge?**

Lactobacilli (95%).

☐ **What causes toxic shock syndrome (TSS)?**

An exotoxin elaborated by certain strains of Staphylococcus aureus. Other organisms causing toxic shock syndrome are group A Streptococcus, Pseudomonas aeruginosa, and Streptococcus pneumoniae. Tampons, IUDs, septic abortions, sponges, soft tissue abscesses, osteomyelitis, nasal packing, and post partum infections can all house these organisms.

OBSTETRICS AND GYNECOLOGY

☐ **What dermatological changes occur with TSS?**

Initially, the patient will have a blanching erythematous rash that lasts for 3 days. 10 days after the start of the infection there will be a desquamation of the palms and soles.

☐ **What criteria are necessary for the diagnosis of TSS?**

All of the following must be present: T > 38.9°C (102°F), rash, systolic BP < 90, and orthostasis, involvement of 3 organ systems (GI, renal, musculoskeletal, mucosal, hepatic, hematologic, or CNS), and negative serologic tests for diseases such as RMSF, hepatitis B, measles, leptospirosis, VDRL, etc.

☐ **What is the treatment for TSS?**

Fluids, pressure support, FFP or transfusions, vaginal irrigation with iodine or saline (if the TSS was related to tampon use), and antistaphylococcal penicillin (nafcillin or oxacillin), or cephalosporin.

☐ **What is the number one cause of UTIs?**

E. coli.

☐ **A 16-year-old female presents to the ED complaining of fever, aches and pains, and painful genital sores that looked like blisters 2 days ago until they popped and started hurting. What would you expect to find on culture of the vesicular fluid?**

This patient most likely has the herpes simplex virus, type 2. Tzanck smear or a culture will show multinucleated giant cells by Giemsa stain.

☐ **What is the most common cause of vaginitis in girls who are pubertal but not sexually active?**

C. albicans.

☐ **Is appendicitis more common during pregnancy?**

No (1/850). However, the outcome is worse. The WBC count usually does not increase beyond the normal value of 12,000–15,000. In a pregnant patient, pyuria with no bacteria suggest appendicitis. Pregnant patients may lack GI distress, and fever may be absent or low grade.

☐ **How is the appendix displaced in pregnancy?**

Superiorly and laterally. Diagnosis of appendicitis in pregnant patients may be further complicated by the fact that normal pregnancy can itself cause an increased WBC. Prompt diagnosis is important because incidence of perforation increases from 10% in the first trimester to 40% in the third.

☐ **A 15-year-old, Tanner 5 girl presents with abdominal pain and dysuria; she has not yet started menses. Examination of the genitalia shows a mass at the introitus. What is the most likely diagnosis and cause?**

The most likely diagnosis is hematocolpos, which is due to either a transverse vaginal membrane or an imperforate hymen.

☐ **A 3-year-old girl presents with painless vaginal bleeding; there is no history of trauma. Examination of the genitalia reveals a donut shaped, erythematous mass above the introitus. What is the diagnosis?**

Urethral prolapse.

☐ **What is the treatment of urethral prolapse?**

If there is no tissue necrosis, management consists of topical estrogen cream for 4-6 weeks. If there is tissue necrosis, excision under general anesthesia is necessary, which can be done as an outpatient.

☐ **A mother presents to the ED because she fears her 7 month old daughter's vagina is "closing up". Examination reveals midline fusion of the labia minora, obscuring the view of the introitus. What is the diagnosis?**

Labial adhesions or acquired attachment of the labia minora. This is thought to occur because of low levels of endogenous estrogen in preadolescent girls, resulting in a thin epithelial surface of the labia minora. This thin epithelial surface is prone to maceration and inflammation, leading to fusion of the edges.

☐ **What is the treatment for labial adhesions?**

Topical estrogen cream for 4-6 weeks, followed by application of an inert cream, such as zinc oxide, for an additional 2 weeks, to keep the healing surfaces apart.

☐ **What is primary dysmenorrhea?**

Painful menses that is not attributable to any other cause, such as endometriosis, PID or structural pelvic disorders.

☐ **What are the treatment options for primary dysmenorrhea?**

First line therapy consists of prostaglandin inhibitors such as NSAIDS. For the sexually active teen, oral contraceptives, which will inhibit ovulation, are another alternative, especially if NSAIDS fail to work.

☐ **What is the pathophysiology behind dysfunctional uterine bleeding (DUB)?**

DUB is due to anovulatory bleeding, caused by estrogen that is unopposed by progesterone. This causes the endometrium to undergo sporadic growth and sloughing without cyclic coordination. The amount of estrogen secreted by the ovaries varies and bleeding can occur because of a fall in estrogen (withdrawal bleeding) or when the endometrium outgrows its blood supply (breakthrough bleeding).

☐ **Among teenagers who are sexually active, what are the 3 most common causes of vaginitis?**

Gardnerella vaginalis (bacterial vaginosis), C. albicans, Trichomonas vaginalis.

☐ **How is the diagnosis of vaginitis by Trichomonas vaginalis made?**

Examination of a wet mount preparation of the vaginal discharge showing flagellated protozoa.

OBSTETRICS AND GYNECOLOGY

☐ **How does an adolescent female acquire vaginitis due to Trichomonas vaginalis?**

Sexual contact.

☐ **When treating vaginitis due to Trichomonas vaginalis, what condition must be excluded before metronidazole is prescribed?**

Pregnancy, as metronidazole is teratogenic.

☐ **With what drugs does metronidazole interact?**

Ethanol, phenytoin, phenobarbital, warfarin and disulfiram.

☐ **A nine-week-old female infant has had a white, malodorous discharge for over seven weeks. Her physical examination is normal except for the aforementioned discharge. A wet prep of the discharge shows motile trichomonads. How did the infant acquire this infection?**

A small number of vaginally delivered female infants may acquire vaginitis due to Trichomonas vaginalis from their untreated mothers during delivery. This vaginal discharge may persist for several months if left untreated.

☐ **A 10-year-old girl presents with vaginal pain, pruritus and a yellow vaginal discharge. She also has a high fever and a sore throat. A gram stain of the vaginal discharge shows gram positive cocci in pairs. What is the most likely diagnosis, and what is the treatment?**

Group A beta hemolytic streptococcus (GABHS). Vaginitis can appear with or following acute GABHS pharyngitis. Penicillin therapy is curative.

☐ **What historical, clinical and laboratory findings are necessary to make the diagnosis of bacterial vaginosis (vaginitis due to Gardnerella vaginalis)?**

A whitish, adherent vaginal discharge; vaginal pH > 4.5; amine-like (fishy) odor generated after adding 10% KOH to a sample of the vaginal discharge; when examining the vaginal discharge under the microscope, > 20% of vaginal epithelial cells are studded with gram negative coccobacilli.

☐ **What are two treatment options for teenage girls with bacterial vaginosis?**

Oral metronidazole twice daily for 7 days. Topical clindamycin is an alternative for patients who are pregnant.

☐ **What is the most common bacteria isolated from a Bartholin's gland abscess?**

N. gonorrhoeae is isolated in up to 50% of cultures from Bartholin's gland abscesses. The remaining cases contain mixed growths of aerobic and anaerobic organisms, such as Bacteroides, and a smaller number will grow C. trachomatis. Up to 30% of cultures will be sterile.

☐ **What is Fitz-Hugh Curtis syndrome?**

Fitz-Hugh Curtis syndrome is a peri-hepatitis, or inflammation of the liver capsule, in association with pelvic inflammatory disease (PID). In addition to the signs/symptoms of PID, patients will also complain of right upper quadrant pain and tenderness to palpation.

PEDIATRIC EMERGENCY MEDICINE BOARD REVIEW

☐ **What other complications are seen in patients with PID and Fitz-Hugh Curtis syndrome?**

20% of patients will have tubo-ovarian abscesses.

☐ **What does "RPR" stand for and what does it detect in serum?**

RPR stands for "rapid plasma regain" and detects antibodies against a cardiolipin-cholesterol-lecithin complex. It is a "nonspecific" test for syphilis, as it does not directly detect Treponema pallidum.

☐ **What is the most likely causes of vaginal discharge in a neonate?**

Physiologic leukorrhea, which is due to the influence of maternal hormones during gestation. The vaginal discharge is clear, white and sticky. A gram stain will show white blood cells but no organisms; cultures will be sterile.

☐ **What pathogen is most often responsible for vaginal discharge immediately beyond the neonatal period?**

Trichomonas vaginalis, which is acquired during vaginal delivery of infants born to infected mothers.

☐ **What is the cause of condyloma acuminata (genital warts)?**

Human papilloma virus.

☐ **A 10-year-old girl presents with chronic vaginal itching; there is no vaginal discharge. On examination her perineum has a hypopigmented, macular rash with telangiectasias in a "figure of 8" pattern. What is the diagnosis?**

Lichen sclerosus, which is a chronic idiopathic dermatitis characterized by atrophy, telangiectasias and hypopigmentation of the perineum, often in a "figure of 8" pattern.

☐ **What is the most common etiology of genital ulcers in the United States?**

Herpes simplex, type 2.

☐ **How can one make the diagnosis of primary syphilis?**

1) Dark field microscopy of suspicious lesions, which will show the spirochete, 2) RPR, which is a good but nonspecific test, as it does not directly detect Treponema pallidum, or 3) fluorescent antibody testing (MHA-TP, FTA), which directly detect antibody response to T. pallidum.

☐ **A teenage girl presents with milky, non-bloody discharge from her nipples. She is not pregnant. What is the diagnosis and etiology of the discharge?**

This patient has galactorrhea, which is most often caused by a pituitary adenoma that secretes prolactin. Galactorrhea can also be caused by certain drugs, such as phenothiazine, antihistamines, diazepam, cimetidine, metoclopramide, opiates (codeine, morphine), antihypertensive agents (verapamil) and tricyclic antidepressants. If amenorrhea is also present, thyroid disease should be suspected.

OBSTETRICS AND GYNECOLOGY

☐ **What is oligomenorrhea?**

Oligomenorrhea is defined as an interval of more than 6 weeks between menses.

☐ **At what point does oligomenorrhea require evaluation?**

Adolescents with a pattern of oligomenorrhea that has continued more than 2 years after menarche or that began after the establishment of a regular menstrual cycle need evaluation.

☐ **What are the clinical manifestations of secondary syphilis?**

Maculopapular rash of the trunk, soles, palms, that can resemble pityriasis rosea; mucous patches (white patches on the mucous membranes); condyloma lata, which are flat topped warts present in moist areas of the body, such as the perineum.

☐ **What is the treatment for sexually acquired syphilis?**

For primary and early (< 1 year duration of symptoms) syphilis, a single dose of IM benzathine penicillin is sufficient; for late (>1 year duration of symptoms) syphilis, 3 IM doses of benzathine penicillin separated by a weekly interval is required.

☐ **What is the most common cause of urethritis among sexually active adolescents in the United States?**

Chlamydia trachomatis.

☐ **What are the clinical manifestations of primary syphilis?**

A painless genital ulcer (chancre), which is the site of inoculation of the spirochete.

☐ **A sexually active 16-year-old male complains of dysuria and a penile discharge. You suspect urethritis. What is the appropriate management of this patient?**

The two most common causes of urethritis in a sexually active male are C. trachomatis and N. gonorrhoeae: therefore, give a third-generation cephalosporin (ceftriaxone) and an oral course of doxycycline.

☐ **Is Candida albicans a common cause of vulvovaginitis in pre-pubescent females?**

No, because estrogen promotes fungal growth by stimulating vaginal glycogen stores and acidity.

☐ **What are some of the risk factors for developing vulvovaginitis from C. albicans?**

Estrogen, which increases vaginal stores of glycogen and increases vaginal acidity; impaired host immunity, due to systemic corticosteroids, diabetes mellitus; and broad spectrum antibiotics.

☐ **What are the signs and symptoms of pelvic inflammatory disease (PID)?**

Fever, lower abdominal pain and marked bilateral adnexal tenderness.

☐ **What method of contraception is associated with a higher incidence of pelvic inflammatory disease (PID)?**

The risk of PID is twice as high in women who use intrauterine devices (IUD) when compared with women who use no contraceptive method.

☐ **Which micro-organisms are responsible for pelvic inflammatory disease (PID)?**

N. gonorrhoeae, C. trachomatis, Mycoplasma hominis, and enteric anaerobes as Bacteroides.

☐ **What is the appropriate antibiotic therapy for the treatment of an adolescent who has cervicitis?**

This patient may have cervicitis from N. gonorrhoeae and/or C. trachomatis. One strategy would be to give a single intramuscular dose of ceftriaxone and a 7-day course of oral doxycycline (provided the patient is not pregnant). To enhance compliance, a single oral dose of azithromycin may be substituted for the doxycycline. To avoid an intramuscular shot, a single oral dose of cefixime may be substituted for ceftriaxone.

☐ **Would the treatment options change if the patient in the previous question were pregnant?**

Yes. If the patient is pregnant, a 7-day course of erythromycin should be substituted for doxycycline.

☐ **A ten-year-old girl presents with a unilateral hematoma of the labia majora after a straddle injury. What is the appropriate treatment for this patient?**

Vulvar hematomas usually respond to ice-packs and bed rest.

☐ **A sexually active teenage girl presents with fever, rash and pain of her knees and elbows. She just completed her menses. Examination reveals fever, a swollen, red, hot knee and erythematous macules of her hands and soles of her feet. What is the most likely diagnosis?**

Disseminated gonococcal infection.

☐ **A teenage girl is treated for secondary syphilis with a single intramuscular dose of benzathine penicillin. Several days later, she develops fever and the maculopapular rash on her trunk becomes much more prominent. What is the cause of her symptoms?**

This patient has an acute systemic febrile reaction, the Jarisch-Herxheimer reaction, which can be seen in up to 20% of patients with acquired or congenital syphilis that are treated with penicillin. This reaction is self-limited and is not a reason to discontinue future therapy with penicillin.

☐ **What are the clinical manifestations of disseminated gonococcal infection?**

Arthritis, polyarthralgia, tenosynovitis and a papular rash. Most patients do not have genitourinary symptoms; however.

OBSTETRICS AND GYNECOLOGY

☐ **What is lymphogranuloma venereum (LGV)?**

An invasive lymphatic infection caused by C. trachomatis.

☐ **What is pseudocyesis?**

A rare cause of amenorrhea in women who believe they are pregnant, when they are not. These patients exhibit many symptoms of pregnancy, such as nausea, vomiting, galactorrhea and abdominal distention. The diagnosis is made when the patient insists she is pregnant, despite the absence of fetal heart tones, fetal movement, uterine enlargement and a negative pregnancy test. All patients require psychiatric evaluation.

☐ **What is the classic triad of symptoms seen in patients with ectopic pregnancies?**

Lower abdominal pain, vaginal bleeding and an abnormal menstrual history.

☐ **What is the most common presentation of lymphogranuloma venereum (LGV)?**

The initial presentation is that of a local lesion on the genitalia accompanied by regional lymphadenitis.

☐ **What is the treatment for lymphogranuloma venereum?**

Tetracycline (patients > 9 years of age), erythromycin, azithromycin or a sulfonamide for 3 to 6 weeks.

☐ **How is the diagnosis of infection with N. gonorrhoeae made?**

By incubating infected secretions on selective media, such as chocolate agar impregnated with antibiotics, in a 5-10% CO_2 environment; this is the gold standard for diagnosing gonococcal infection.

☐ **How soon, after conception, do serum pregnancy tests become positive?**

Within 7 days of conception, which is approximately 1 week before the first missed menstrual period.

☐ **What is the recommended treatment for the first episode of genital herpes?**

If oral acyclovir is started within 6 days of the onset of lesions, the duration of the lesions may be shortened by 3-5 days.

☐ **What is the recommended treatment for recurrent genital herpes?**

Oral acyclovir has minimal effect on the duration of recurrent episodes of genital herpes. If oral acyclovir is started within 48 hours of the onset of symptoms, the duration of the lesions may be shortened only by 1 day.

☐ **A 13-year-old girl, Tanner stage 4, presents with yellow vaginal discharge. She has not started menstruating, and is not sexually active. There is no vaginal itching, pain or dysuria. What is the most likely diagnosis?**

Physiologic vaginal discharge, which is usually scant, yellow and without odor; there are no other associated symptoms. This occurs about 6 months prior to menarche.

☐ **What are the typical symptoms of "mittelschmerz" (ovulatory pain)?**

1. Sudden onset of right or left lower quadrant pain that occurs 2 weeks after the last menses (mid-cycle).
2. No evidence of intrauterine pregnancy.
3. A pelvic examination that reveals unilateral adnexal tenderness with no mass, no cervical motion tenderness, and no vaginal bleeding.
4. Pain that lasts only 24-36 hours.

☐ **In an adolescent female without a history of PID, what is the most common cause of chronic pelvic discomfort?**

Endometriosis.

☐ **When is the pain of endometriosis most severe?**

Right before menses.

☐ **How soon after conception is the average home pregnancy kit able to detect the HCG and register as positive?**

Depending on sensitivity, anywhere from 1 to 3 weeks.

☐ **In a normal uterine pregnancy, what is the doubling time of the HCG levels?**

48 hours. This can be used as one of several screening measures to rule out a hydatidiform mole.

LEGAL

☐ **In general, do minors have the right to give consent for their own treatment?**

No--this right belongs to their parents or guardians.

☐ **In an emergency, is it necessary to obtain consent in order to treat a child?**

No.

☐ **May a child who lives away from home, is no longer subject to parental control, is economically self-supporting, is married, or is a member of the armed forces give or refuse consent to treatment?**

In general, yes. Such children are considered emancipated minors.

☐ **May a child who is sufficiently mature to understand the nature of his/her illness and the potential risks and benefits of proposed therapy give or refuse consent to treatment?**

In general, yes. Such children are considered mature minors.

☐ **Are there other circumstances in which a child has the right to give informed consent to treatment?**

Yes--many states have laws giving children such rights for treatment of sexually transmitted or other communicable diseases, crisis counseling, or substance abuse.

☐ **What are the physician's legal obligations where a minor child requests that information obtained in the course of treatment not be communicated to the child's parents?**

In the absence of circumstances such as those outlined in the preceding questions, the physician may not legally honor such a request.

☐ **May Jehovah's Witness parents refuse blood transfusions for their minor children?**

No.

☐ **May parents be criminally prosecuted if their child dies because they treated the child with prayer rather than with medicine?**

Yes, in some jurisdictions.

☐ **May parents authorize removal of life support for a minor child in a persistent vegetative state?**

Yes.

☐ **Is it legally required for a physician who suspects child abuse or neglect to report it?**

Yes.

☐ **Is reporting required only for suspected physical abuse?**

No--emotional abuse, sexual abuse, and other forms of abuse are covered by reporting statutes.

☐ **May a parent's refusal to authorize indicated medical treatment be considered child neglect for purposes of the reporting statute?**

Yes.

☐ **May a physician suffer a penalty for failure to report suspected abuse, neglect, or endangerment?**

Yes.

☐ **May a physician be sued by a child or family for failing to report suspected abuse, neglect, or endangerment?**

Yes.

☐ **What would it take for the child or family to win such a suit?**

They must show that the failure to report contributed significantly to a subsequent injury that might have been averted.

☐ **May a physician be sued for defamation based on her/his report of suspected abuse, neglect, or endangerment?**

No, if the report was made in good faith.

☐ **May mothers who engage in substance abuse while pregnant be the subject of criminal prosecution?**

No--such prosecutions have been brought but not upheld.

☐ **In order for a physician to be guilty of malpractice, must the existence of a physician-patient relationship be proven?**

Yes.

LEGAL

☐ **What is medical negligence?**

Acting below the accepted standard of care.

☐ **Is an emergency physician legally required to do what the best specialist in the field would do?**

No--only what a typical reasonable competent physician would do under the circumstances.

☐ **In order for a physician to be found guilty of malpractice, must it be shown that the patient was injured?**

Yes.

☐ **How are practitioners and hospitals that provide care for children legally judged?**

By widely accepted pediatric standards of care.

☐ **In order for a physician to be found guilty of malpractice, must it be shown that the patient's injury was caused by the physician's act or omission?**

Yes.

☐ **Are there any exceptions to the "informed consent" doctrine?**

Yes, consent may be implied if needed treatment is emergent or the patient's ability to make a decision is lost.

☐ **Do all states follow the same common-law definition of "informed consent"?**

No, you must strictly adhere to your state's regulations.

☐ **Does the law protect the parents in fully determining the care of their child?**

Yes, if the parents act in the best interest of the child.

☐ **T/F: "Baby Doe" cases involve ethical decisions by providers refusing to treat a child with an infectious disease.**

False. "Baby Doe" cases include decision making in cases with severely handicapped or critically ill newborns. Decisions are made by the parents and physicians involved in the child's care.

☐ **Identify one resource the provider can utilize in dealing with complex decisions surrounding the care of neonates?**

The hospital ethics committee is a valuable resource for the healthcare providers.

PEDIATRIC EMERGENCY MEDICINE BOARD REVIEW

☐ **How can a child acquire status of an "emancipated minor"?**

Marriage, judicial decree, military service, parental consent, failure of parents to meet legal responsibilities, living apart from and being financially independent of parents, motherhood.

☐ **T/F: Under the term "vicarious liability," the team leader is responsible for the actions of others.**

False. Each team member is held responsible for his or her actions. Each team member is held to the standard of care for his or her level of training.

☐ **What specific criteria for brain death in children have been established and validated?**

None. There are a number of definitions used to define brain death in children. Circumstances such as hypothermia, drug use, or idiopathic causes of brain injury or insult may alter neurologic criteria determining the diagnosis of death.

☐ **Is organ donation a consideration when a child is the victim?**

Yes. If approached with respect and sensitivity, organ donation may ultimately benefit the child's family.

☐ **T/F: If the healthcare provider can safely care for the adult patient, he or she can safely care for a child.**

False. Children are different from adults. The provider must have a sound knowledge base with pediatric standards of care.

☐ **When does the legal duty to meet a "reasonable standard" of care commence for the healthcare provider?**

Once he or she performs an act that may be construed as rendering care.

☐ **What form of consent traditionally applies to the emergency situation?**

Implied consent.

☐ **If a healthcare provider treats a patient without first being granted permission, what may that provider be guilty of committing?**

Assault.

☐ **What are the three essential operational elements of informed consent?**

1. A careful assessment of the patient's decision-making capacity.
2. A judgment as to the "voluntariness" of the decisions.
3. A determination of the nature and extent of the information to be disclosed.

LEGAL

☐ **In what three ways is patient decision-making capacity reflected?**

In his or her ability to comprehend, communicate, and appreciate the consequences of available choices.

☐ **To what does "voluntariness" refer?**

The absence of internal or extrinsic pressures that may coercively restrict patient options.

☐ **What does the "reasonable person" standard require physicians to disclose when obtaining informed consent from a patient?**

Those material facts concerning treatment alternatives and risks that a reasonable person would need to make an informed decision.

☐ **What is "therapeutic privilege?"**

If there is strong reason to believe that the disclosure involved in obtaining informed consent may cause serious psychological harm, physicians may be excused for failing to obtain it.

☐ **T/F: States vary considerably with respect to their statutory or common-law definition of informed consent.**

True.

☐ **When may parents make medical decisions on behalf of their children contrary to physician advice?**

On condition they fulfill the duty to provide necessary care for minor children.

☐ **When may the state assert its interests in protecting the welfare of children by invoking child protection statutes to override parental wishes?**

If parents fail to provide their children with at least a minimum standard of medical care.

☐ **If a parent makes a decision on behalf of his or her child motivated by strong family convictions, can the courts interfere?**

Courts regularly uphold such interventions when parent refusals may be life threatening.

☐ **What if the consequences are grave, but not life threatening?**

States have varied in their willingness to intervene in such cases, reflecting the continuing struggle to balance the rights of individual children and family privacy.

☐ **Are there exemptions to child protection statutes for parents who seek non-traditional forms of treatment based on religious convictions?**

Many state child-protection statutes have included such exemptions, although the American Academy of Pediatrics has urged states to repeal such provisions.

☐ **According to Congressional guidelines, under what three conditions may treatment be withheld from neonates?**

1. The infant is chronically and irreversibly comatose.
2. Treatment would merely prolong dying or would not be effective in ameliorating or correcting all of the infant's life-threatening conditions.
3. Treatment would be virtually futile in terms of survival and therefore inhumane.

☐ **How is a "minor" defined?**

Historically, the right of self-determination is recognized at the legal age of maturity, 18 years. Below this age a patient is considered a minor.

☐ **Do minors have the right to consent to medical treatment?**

Many state legislatures and courts have expanded minors' rights to consent to medical treatment, although great variability in this approach exists from state to state.

☐ **What are "mature minor" rules?**

Statutory rules that uphold the validity of consent given by minors if the treatment is appropriate and the minor is considered capable of comprehending the clinical circumstances and therapeutic options.

☐ **What are two conditions that some states allow minors to consent to treat?**

Venereal disease and substance abuse.

☐ **What is a "variable competence approach" to minors' consent?**

One that considers developmental aspects of cognitive and psychosocial maturation.

☐ **In order to prevail in a malpractice suit, what must the patient prove?**

That the healthcare provider failed in his or her duty to provide the patient with the degree of knowledge, skill, and care usually exercised by a reasonable and prudent provider under similar circumstances, given the prevailing state of medical knowledge and available resources, and that the specific injury was caused by that failure.

☐ **Are errors in judgment evidence of malpractice?**

Not by themselves.

☐ **When does a provider's legal duty toward a patient begin?**

With a mutual agreement to provide care for compensation or with any act that may represent an undertaking to provide care.

LEGAL

☐ **Must a patient have suffered an injury in order to prevail in a malpractice suit?**

Yes. Even if evidence demonstrates that the care provided was substandard, a claim of medical malpractice will fail if the patient escaped injury.

☐ **What is "standard of care"?**

The degree of care and skill expected of a reasonably competent practitioner of the same class, acting in the same or similar circumstances.

☐ **T/F: Failure to obtain consent can be considered malpractice.**

True. Proof of negligence in malpractice cases may be based on failure to obtain informed consent.

☐ **What three things must be proven in order to prove malpractice based on lack of consent?**

1. That the provider failed to disclose all the material facts that a reasonable person would require to make an informed decision.
2. That the patient's injury was caused by the provider's act, for which the patient granted uninformed permission.
3. That a reasonable person would have withheld consent had the material facts been disclosed.

☐ **Since consent in CPR may be implied in cardiac arrest cases, when may problems arise?**

In emergency situations when documents on the victim's person or statements made by family members purport to represent the victim's intended refusal of CPR.

☐ **What is the generally recognized practice in such situations?**

To initiate resuscitative efforts, with the understanding that CPR may be discontinued based on subsequent authentication of the patient's previously expressed wishes.

☐ **What is the "captain-of-the-ship" doctrine?**

That a leader be held vicariously liable for the conduct of others in the group, depending on his or her level of control or right of control of the group members.

☐ **How are conflicts between legitimate parental rights of family privacy and the state's interest in protecting the welfare of children decided?**

On whether life-sustaining treatment serves the best interest of an individual child.

☐ **T/F: Criteria developed for brain death in adults and children are the same.**

False. Specific criteria for brain death developed for adults have not been validated for children.

☐ **What two important elements should hospital policies governing orders not to resuscitate contain?**

1. Requirements that do-not-resuscitate orders be written on the order sheet, accompanied by progress notes that explain the rationale for the decision and identify the participants in the decision-making process.
2. Guidelines for prior judicial review in specifically enumerated circumstances.

☐ **What one of two conditions must be met to conform to the Model Uniform Determination of Death Act?**

1. Irreversible cessation of circulation and respiratory function OR
2. Irreversible cessation of all functions of the entire brain including the brain stem

☐ **T/F: Cardiopulmonary support systems may be withdrawn from brain-dead patients without judicial review or fear of legal repercussions.**

True.

BIBLIOGRAPHY

BOOKS/ARTICLES:

Advanced Cardiac Life Support. Dallas: American Heart Association, 2015.

American Academy of Pediatrics. Neonatal Resuscitation. 7th ed. 2016.

Advanced Trauma Life Support. Chicago: American College of Surgeons, 2012.

American Academy of Ophthalmology. Basic and Clinical Science Course. Section 8 (1994-95), Section 9 (1996-97) and Section 12 (1997-98).

Agur, M. R. Grant's Atlas of Anatomy (13th Ed.). Baltimore: Williams & Wilkins, 2012.

Auerbach, P.S. Management of Wilderness and Environmental Emergencies (6th Ed.). St. Louis: C.V. Mosby Company, 2011.

Bakerman, S. ABCs of Interpretive Laboratory Data (4th Ed.). Greenville: Interpretive Laboratory Data, Inc., 2002.

Barkin, R.M. Emergency Pediatrics (4th Ed.). St. Louis: C.V. Mosby Company, 2004.

Bork, K. Diagnosis and Treatment of Common Skin Diseases (2nd Ed). Philadelphia: W.B. Saunders Company, 1999.

Chameides, Leon, et al. eds. Pediatric Advanced Life Support Provider Manual. Dallas, American Heart Association, 2011.

Christos S., Gossman W. Mnemonics and Pearls Handbook. 2016 edition. Chicago, IL: Wenckebach Publishing, 2016.

Dambro, M.R. Griffith's 5 Minute Clinical Consult. Lippincott, Williams and Wilkins, 2006.

Ellenhorn M.J., Schonwald S., Ordog G., Wasserberger J., Ellenhorn's Medical Toxicology (3rd Edition), Williams & Wilkins, 2003.

Fitzpatrick, T.B. Color Atlas and Synopsis of Clinical Dermatology (4th Ed). New York: McGraw-Hill Publishing Company, 2000.

Gerstenblith, A., Rabinowitz, M. (eds.) The Wills Eye Manual: Office and Emergency Room Diagnosis and Treatment of Eye Disease, Second Edition. JB Lippincott Co., Philadelphia. 2012.

Goldrank, L.R., Flomenbaum N.E., Lewin N.A., Weisman R.S., Howland M.A., Hoffman R.S., Goldfrank's Toxicologic Emergencies (10th Edition), Appleton & Lange 2014, Stamford, CT.

Harris, J.H. The Radiology of Emergency Medicine (5th Ed.). Baltimore: Lippincott, Williams and Wilkins, 2013.

Harwood-Nuss, A. The Clinical Practice of Emergency Medicine (6rd Ed). Philadelphia: J.B. Lippincott, Williams & Wilkins, 2014.

Haxhija EQ, Nores H, Schober P, Hollwarth ME. Lung contusion-lacerations after blunt thoracic trauma in children. Pediatr Surg Int. 2004 Jun;20(6):412-4.

Hoppenfeld, S. Physical Examination of the Spine and Extremities. Norwalk: Appleton-Century-Crofts, 1976.

Moore, K.L. Clinically Oriented Anatomy. Baltimore: Williams & Wilkins, 2013.

Nelson, Waldo E. Textbook of Pediatrics. Philadelphia: W.B. Saunders Comany, 2004.

Physicians' Desk Reference (58th Ed.). Oradell: Medical Economics Company Inc., 2004.

Plantz, S.H. Emergency Medicine Pearls of Wisdom. (6th Ed.). Lincoln, NE: Boston Medical Publishing, 2005.

Porter, R. S. The Merck Manual. (19th Ed.). 2011.

Rivers, C.S. Preparing for the Written Board Exam in Emergency Medicine Ohio Chapter, ACEp. 6th Ed. 2011.

Rosen, P. Emergency Medicine Concepts and Clinical Practice (20th Ed.). Philadelphia: W.B. Saunders Compan, 2013.

Rowe, R.C. The Harriet Lane Handbook (16th Ed.). Chicago: Year Book Medical Publishers, Inc., 2014.

Simon, R.R. Emergency Orthopedics The Extremities (6th Ed.). Norwalk: Appleton & Lange, 2011.

Simon, R.R. Emergency Procedures and Techniques (7th Ed.). Baltimore: Lippincott, Williams and Wilkins, 2014.

Squire, L.F. Fundamentals of Radiology (5th Ed.). Cambridge: Harvard University Press, 1997.

Strange, G. Pediatric Emergency Medicine: Just the Facts. (2nd Ed). McGraw-Hill. 2011.

Tintinalli, J.E. Emergency Medicine A Comprehensive Study Guide (8th Ed.). New York: McGraw-Hill, Inc., 2015.

Weinberg, S. Color Atlas of Pediatric Dermatology (4th Ed.). New York: McGraw-Hill, 2001.

Made in the USA
Lexington, KY
20 July 2018